SCIENTIFIC ADVANCES IN ANIMAL NUTRITION

Promise for the New Century

Proceedings of a Symposium

Committee on Animal Nutrition

Board on Agriculture and Natural Resources

Division on Earth and Life Studies

National Research Council

NATIONAL ACADEMY PRESS
Washington, D.C.

NATIONAL ACADEMY PRESS · 2101 Constitution Avenue, NW · Washington, D.C. 20418

NOTICE: The project that is the subject of this report was approved by the Governing Board of the National Research Council, whose members are drawn from the councils of the National Academy of Sciences, the National Academy of Engineering, and the Institute of Medicine. The members of the committee responsible for the report were chosen for their special competences and with regard for appropriate balance.

This material is based upon work supported by the National Research Council. Any opinions, findings, conclusions, or recommendations expressed in this publication are those of the authors and do not necessarily reflect the view of the Research Council.

ISBN 0-309-08276-5

Copyright 2001 by the National Academy of Sciences. All rights reserved.

Printed in the United States of America.

THE NATIONAL ACADEMIES

National Academy of Sciences
National Academy of Engineering
Institute of Medicine
National Research Council

The **National Academy of Sciences** is a private, nonprofit, self-perpetuating society of distinguished scholars engaged in scientific and engineering research, dedicated to the furtherance of science and technology and to their use for the general welfare. Upon the authority of the charter granted to it by the Congress in 1863, the Academy has a mandate that requires it to advise the federal government on scientific and technical matters. Dr. Bruce M. Alberts is president of the National Academy of Sciences.

The **National Academy of Engineering** was established in 1964, under the charter of the National Academy of Sciences, as a parallel organization of outstanding engineers. It is autonomous in its administration and in the selection of its members, sharing with the National Academy of Sciences the responsibility for advising the federal government. The National Academy of Engineering also sponsors engineering programs aimed at meeting national needs, encourages education and research, and recognizes the superior achievements of engineers. Dr. Wm. A. Wulf is president of the National Academy of Engineering.

The **Institute of Medicine** was established in 1970 by the National Academy of Sciences to secure the services of eminent members of appropriate professions in the examination of policy matters pertaining to the health of the public. The Institute acts under the responsibility given to the National Academy of Sciences by its congressional charter to be an adviser to the federal government and, upon its own initiative, to identify issues of medical care, research, and education. Dr. Kenneth I. Shine is president of the Institute of Medicine.

The **National Research Council** was organized by the National Academy of Sciences in 1916 to associate the broad community of science and technology with the Academy's purposes of furthering knowledge and advising the federal government. Functioning in accordance with general policies determined by the Academy, the Council has become the principal operating agency of both the National Academy of Sciences and the National Academy of Engineering in providing services to the government, the public, and the scientific and engineering communities. The Council is administered jointly by both Academies and the Institute of Medicine. Dr. Bruce M. Alberts and Dr. Wm. A. Wulf are chairman and vice chairman, respectively, of the National Research Council.

COMMITTEE ON ANIMAL NUTRITION

GARY L. CROMWELL, *Chair*, University of Kentucky
MARY E. ALLEN, National Zoological Park
MICHAEL L. GALYEAN, Texas Tech University
RONALD W. HARDY, University of Idaho
BRIAN W. MCBRIDE, University of Guelph
KEITH E. RINEHART, Perdue Farms Incorporated
L. LEE SOUTHERN, Louisiana State University
JERRY W. SPEARS, North Carolina State University
DONALD R. TOPLIFF, West Texas A&M University
WILLIAM P. WEISS, The Ohio State University

Staff

CHARLOTTE KIRK BAER, *Program Director*
JULIE BERRY, *Science Writer*
GRETCHEN OPPER, *Policy Intern*
STEPHANIE PADGHAM, *Project Assistant*
MELINDA SIMONS, *Project Assistant**

* *through January 1999*

BOARD ON AGRICULTURE AND NATURAL RESOURCES

HARLEY W. MOON, *Chair*, Iowa State University
CORNELIA B. FLORA, Iowa State University
ROBERT B. FRIDLEY, University of California
BARBARA GLENN, Federation of Animal Science Societies
LINDA GOLODNER, National Consumers League
W.R. (REG) GOMES, University of California
PERRY R. HAGENSTEIN, Institute for Forest Analysis, Planning, and Policy, Wayland, Massachusetts
GEORGE R. HALLBERG, The Cadmus Group, Inc., Waltham, Massachusetts
CALESTOUS JUMA, Harvard University
GILBERT A. LEVEILLE, McNeil Consumer Healthcare, Denville, New Jersey
WHITNEY MACMILLAN, Cargill, Inc., Minneapolis, Minnesota (retired)
TERRY MEDLEY, DuPont Biosolutions Enterprise
WILLIAM L. OGREN, U.S. Department of Agriculture (retired)
ALICE PELL, Cornell University
NANCY J. RACHMAN, Novigen Sciences, Inc.
G. EDWARD SCHUH, University of Minnesota
BRIAN STASKAWICZ, University of California, Berkeley
JOHN W. SUTTIE, University of Wisconsin
JAMES TUMLINSON, USDA, ARS
JAMES J. ZUICHES, Washington State University

Staff

CHARLOTTE KIRK BAER, *Director*
HEATHER CHRISTIANSEN, *Research Associate*

Preface

The science of animal nutrition has made significant advances in the past century. In looking back at the discoveries of the 20th century, we can appreciate the tremendous impact that animal nutrition has had on our lives. From the discovery of vitamins and the sweeping shift in the use of oilseeds to replace animal products as dietary protein sources for animals during the war times of the 1900s—to our integral understanding of nutrients as regulators of gene expression today—animal nutrition has been the cornerstone for scientific advances in many areas.

At the milestone of our 70th year of service to the nation, the National Research Council's (NRC) Committee on Animal Nutrition (CAN) sought to gain a better understanding of the magnitude of recent discoveries and directions in animal nutrition for the new century we are embarking upon. With financial support from the NRC, we were able to organize and host a symposium that featured scientists from many backgrounds who were asked to share their ideas with us about the potential of animal nutrition to address current problems and future challenges.

From this gathering, it became overwhelmingly evident that the answers to many of the important issues facing agriculture and the global population lie in animal nutrition. With few exceptions, animal nutrition accomplishments and opportunities impact almost every aspect of our universe.

We structured the symposium and prepared this proceedings for a diverse audience, from nonscientists and policymakers, to scientists in nutrition and related areas. We hope to reach decision-makers in government, academic institutions, foundations, and private industry, who are in positions to understand, direct, and support continued positive impacts and growth in animal nutrition.

It is our hope that this proceedings will also guide students who are considering professions in animal nutrition, to consider the vast array of potential careers. By compiling this wide-ranging mix of nutrition topics as they impact various world-wide endeavors, we hope to convey our excitement about future opportunities and provide young scientists with an appreciation for the diversity of problems that can be addressed through animal nutrition.

The symposium was organized under the guidance of former chair of the Board on Agriculture and Natural Resources, Dale E. Bauman, Cornell University; former chair of the Committee on Animal Nutrition, Donald C. Beitz, Iowa State University; and members of the Committee on Animal Nutrition. This volume is comprised of individually authored papers that follow, and they fall into five broad categories:

- Conservation aspects of animal nutrition;

- Recent developments in animal nutrition, health, and well-being;

- Animal nutrition's role in endeavors throughout our universe;

- International and economic aspects of animal nutrition; and

- Meeting challenges of the new century.

The papers presented here provide a cohesive view of not only the scientific aspects of animal nutrition, but also speak to the practical application of science in day-to-day living. Many of the papers describe an inherent reliance on the work of the NRC's CAN, which is understandable, given the fundamental role CAN plays in the work of individual scientists, regulators, consultants, professionals, and students. While the information in the proceedings is attributed to the individual perspectives of the authors who shared their expertise, it reflects the collective contributions of hundreds of individuals and groups who have devoted their lives to animal nutrition for the good of the public and the animals for which we are responsible. As a committee, we have benefitted greatly from the ideas presented here and hope that, as a result, this report represents a valuable compilation that puts our thinking for the future into its broadest perspective.

Gary L. Cromwell, *Chair*
Committee on Animal Nutrition

Acknowledgments

The members of the Committee on Animal Nutrition express our thanks to the individuals who participated in this symposium by providing stimulating presentations and who provided us with thoughtful text to support their discussion. We are grateful to the symposium session moderators: Donald C. Beitz, Iowa State University; Mary E. Allen, Smithsonian Institution's National Zoological Park; and Michael Galyean, Texas Tech University. We also wish to thank the session rapporteurs: John Halver, University of Washington; Joseph Fontentot, Virginia Polytechnic and State University; and Karin Wittenberg, University of Manitoba.

The committee appreciates the assistance of science writer, Julie Berry, and National Research Council (NRC) policy intern, Gretchen Opper, in the preparation of this volume. In addition, we thank the many scientists who have provided input and ideas for this symposium and publication.

Our heartfelt thanks and appreciation are extended to our NRC program director, Charlotte Kirk Baer, who enthusiastically championed this commemorative event and worked diligently to bring this work to a successful conclusion. Staff members who also supported us in this endeavor include Stephanie Padgham and Melinda Simons. For their work, we are grateful.

This report has been reviewed in draft form by individuals chosen for their diverse perspectives and technical expertise, in accordance with procedures approved by the NRC's Report Review Committee. The purpose of this independent review is to provide candid and critical comments that will assist the institution in making its published report as sound as possible and to ensure that the report meets institutional standards for objectivity, evidence, and responsiveness to the study charge. The review comments and draft manuscript remain confidential to protect the integrity of the process. We wish to thank the following individuals for their review of this report: Donald Beitz, Iowa State University; Susan Crissey, Brookfield Zoo; Austin Lewis, University of Nebraska; and Robert Wilson, Mississippi State University.

Although the reviewers listed above have provided many constructive comments and suggestions, they did not see the final draft of the report before its release. The review of this report was overseen by Delbert Gatlin, Texas A & M University. Appointed by the National Research Council, he was responsible for making certain that an independent examination of this report was carried out in accordance with institutional procedures and that all review comments were carefully considered. Responsibility for the final content of this report rests entirely with the authoring committee and the institution.

Contents

KEYNOTE ADDRESS: INROADS TO ANIMAL CONSERVATION 1
Jane Goodall

1   LANDMARK AND HISTORIC CONTRIBUTIONS OF ANIMAL NUTRITION 7
Duane E. Ullrey

2   PROTECTING ANIMAL HEALTH AND WELL-BEING: NUTRITION AND IMMUNE FUNCTION 13
Kirk Klasing

3   DESIGNING FOODS: FEEDING ANIMALS TO REDUCE HUMAN HEALTH RISKS 21
Bruce Watkins

4   METABOLIC MODIFIERS: ADVANCES IN ECONOMIC PRODUCTION OF SAFE FOOD 26
Robert J. Collier

5   NUTRIENTS AS REGULATORS OF GENE EXPRESSION 32
Donald Jump

6   OUR CHANGING ENVIRONMENT: DEVELOPING STRATEGIES FOR THE FUTURE 38
Danny Fox

7   READINESS OF MILITARY SERVICE ANIMALS 44
Susan Yanoff and Michelle Ross

8   RESEARCH AND EDUCATION NEEDS FOR THE NEXT GENERATION 48
Quinton Rogers

9   INTERNATIONAL RELEVANCE OF FEED COMPOSITION INFORMATION 53
Philip Thacker

10	THE INTERNATIONAL AQUACULTURE MARKET AND GLOBAL NEEDS .... 57 *Daniel Villamar*	
11	MEETING THE CHALLENGES OF THE NEW CENTURY .... 72 *Dale E. Bauman*	

AUTHORS 78

APPENDIX 85

TABLES AND FIGURES

Tables

2-1   Mechanisms by which nutrition modulates immunocompetence and disease resistance, 16
3-1   Status of designed foods, 23
9-1   Variations in the protein content of common feed ingredients, 55
10–1 Some estimated losses to disease in world shrimp farming, 61

Figures

1-1a Galahad eating the pith of an oil nut palm frond, 3
1-1b Wilke consuming leaves, 4
2-1   Dietary requirements set by the National Research Council are usually based on concentrations that maximize growth and reproduction and prevent known deficiency pathologies, 14
4-1   Phase I: Digestive process, 27
4-2   Phase II: Post-absorptive nutrient use, 28
5-1   Overview of nutrients as regulators of gene expression, 33
5-2   Cholesterol regulation of cholesterol metabolism, 34
5-3   Eicosanoid regulation of gene expression, 36
8-1   Total number of CAN publications and those in the Nutrient Requirement series per 5-year period, 49
8-2   Number of pages in the National Research Council Nutrient Requirements series for reports on swine, dogs, and cats during the last 45 years, 49
8-3   Number of references in the National Research Council Nutrient Requirements series for reports on swine, dogs, and cats during the last 45 years, 49

10-1 Millions of metric ton (t) of capture and aquaculture fisheries production for human consumption, 58
10-2 Top species by unit value, 59
10-3 Severe effects of disease on shrimp production in Thailand and Ecuador, 60
10-4 Intensive, semi-extensive/semi-intensive and extensive aquaculture systems, 62
10-5 Extensive production system: use of animal manure for fertilization, 63
10-6 Extensive production system: harvesting food for fish, 64
10-7 Extensive production system: food fed to fish, 64
10-8 Semi-extensive production system for carp in ponds, 65
10-9 Semi-extensive production system: locally made food mixture, 66
10-10 Intensive production system: phase feeding, 68
10-11 Intensive production system: quality product harvesting, 68
10-12 Intensive production system: aeration, 69

Keynote Address

Inroads to Animal Conservation

JANE GOODALL
The Jane Goodall Institute

"A chimpanzee being held in a Brazzaville zoo was emaciated, hairless, and half blind because he was not receiving adequate nutrition. I put together a group of expatriates, employed some keepers, and built him a small patio in an attempt to improve his life. He put on weight, grew hair, and regained his eyesight. He was thrilled to have branches that allowed him to nibble the leaves and make nests. And for the first time since 1944, when he was put in his cage, he was able to go outside and feel the warmth of the sun. He is now one of the oldest chimpanzees in captivity. I do not know how he survived all those years without proper food. Some indomitable spirit in him kept him alive."

Louis Leakey believed that he could learn more about the probable behavior of our earliest Stone Age ancestors from a study of the chimpanzees in the wild. He argued that behavior common to modern chimpanzees and humans was probably also present in a common ancestor millions of years ago. Leakey is well vindicated, because most human evolution textbooks describe the chimpanzee, and specifically the Gombe chimpanzee, behavior in some measure.

Scientific attitude has changed towards these relatives of ours, the nonhuman primates. In the 1960s, the strict ethnological science of Europe did not believe that animals had personalities. Only humans had personalities. Animals were presumed to have no ability for rational thought and problem solving. The worst sin of anthropomorphism was that animals be credited with any kind of emotion.

It is fascinating that since 1960, attitudes have softened and there is no longer a passion for reductionism. People are much more prepared to look at the societies of nonhuman animals and see complexity and individuality. Discussing emotions is usually acceptable if it is done in the right way. The animal mind is now a popular study of many graduate students.

CHARACTERISTICS OF THE GOMBE CHIMPANZEES

The Gombe stream is a 30-square-mile area. It stretches for 10 miles along the shore of Lake Tanganyika, a steep, hilly country, falling down forested slopes from the Rift Escarpment. That area is home to approximately 100 chimpanzees, who have provided us with a wealth of information about primate behavior, including feeding behavior and diet selection, among many others.

The main study community of Gombe consists of some 50 individuals — adults, adolescents, and infants. Male chimpanzees are more overtly aggressive and fight more than females. But because they are ordered in a dominance hierarchy, where males know their positions relative to each other and are dominant to all females, disputes within a community can often be settled by a threatening posture or gesture. A male will bristle his hair, bunch his lips in a ferocious scowl, swagger, brandish sticks and so on.

After some kind of aggression, the victim even though fearful of the more dominant aggressor is likely to approach with some kind of submissive gesture, such as a crouch. In response, the aggressor is likely to reach out with a reassurance behavior such as patting, touching, or even kissing and embracing. And so, social harmony is quickly restored to the group, even after quite serious aggression.

Nonverbal communication patterns of the chimpanzees almost uncannily resemble some human postures and gestures and tend to occur in the same type of context. A nervous female may reach her hand out for reassurance and the male may gently calm her by patting her hand. An adult male may be greeted with a kiss when he joins a young female. Friendly physical contact in chimpanzee society maintains friendships and improves bad relationships. Males will spend long hours peacefully grooming each other, but if two chimpanzees do not like each other, they will not groom. Many of these patterns are inborn; but a young chimpanzee raised in social isolation although he may use these postures and gestures, will do so in inappropriate contexts.

A wild female chimpanzee gives birth approximately every 5 years and usually has her first baby when she is between 11 and 13 years of age. Age of first birth is directly related to body weight, which may be correlated with nutrition. Young chimpanzees in captivity, consistently fed a nutritious diet, can have babies as young as 7 or even 6 years of age. Usually one baby is born, but twins do occur.

Childhood is a time of much activity and a great deal of that activity is play. Young chimpanzees are very well tolerated by other non-related adults in the community. Infants up to 4 years old can take great liberties with their elders. As they move toward adolescence, they become more cautious, particularly in their dealings with adult high-ranking males. Young males must be especially cautious.

NATURAL DIET AND FEEDING BEHAVIOR

Food is plentiful for most of the year in this part of Africa. Chimpanzees are omnivores but the greatest proportion of the wild chimpanzee diet is fruits, which change in type, quantity, and nutritional quality according to the season. Many of the big forest trees have a pattern of fruiting every second year. But sometimes trees fail to fruit and then the chimpanzees lose weight apparently due to decreased food supply.

Chimpanzees in the wild spend a certain amount of time feeding on one type of food, and then, they will usually move on and feed on something else. The variety of their diet is impressive — over 600 different foods are eaten in some areas. As well as fruit, they eat leaves, nuts, shoots, stems, bark, blossoms, seeds, insects, bird eggs and meat (Figures 1a,b). Based on observations of wild chimpanzees, it is clearly preferable for the diets of captive animals to be varied, so they do not get bored with their food.

FIGURE 1-1a. Galahad eating the pith of an oil nut palm frond. (Photograph by Jane Goodall)

FIGURE 1-1b. Wilke consuming leaves. (Photograph by Jane Goodall)

Chimpanzees are good hunters, and, after a successful kill, other chimpanzees gather around the hunter and beg for a share. Sharing may or may not take place depending on the personality of the hunter, the amount of meat and his relationship with those who are begging. Usually after a fairly big animal has been killed, high-ranking males rush in and come away with some portion of the carcass.

Although meat is a highly preferred food and "kills" stir up excitement, meat comprises only 2 percent of the chimpanzee's natural diet. Primates are the preferred and most frequently caught prey for chimpanzees across Africa. Colobus monkeys are the most frequently killed at Gombe. Young pigs and young bush buck are also hunted.

Insects are eaten much more frequently, and termites are popular at Gombe. Termites that fly off from the nest to form new colonies are caught by hand. At other times, chimpanzees will scrape open a tunnel into a termite mound and use a piece of grass to draw them out. Sometimes they strip leaves from a twig for this purpose. It was these observations, made in 1960 because that prompted the National Geographic Society to start funding research, at that time it was thought that only humans used and made tools. Females may "fish" for termites for up to five hours to consume the protein and lipid rich termites. Males seldom termite fish for more than an hour and do not termite fish as frequently as females. They capitalize on the termite season, when the rains begin and when every tool is likely to yield an abundance of insects.

Chimpanzees also eat army ants, which bite fiercely. They live in underground nests. A chimpanzee approaches, opens the entrance by digging in the earth with one hand. He or she then selects a long, thin and very straight stick, peels off the bark and any twigs to make a smooth tool. Then, often sitting on some branch off the ground, he pushes the stick into the nest, waits for a mass of ants to swarm up, sweeps them off into one hand, and crunches them up as fast as possible. After a few minutes, he usually runs off to slap and pick off the ants that have started biting his legs and arms.

Different feeding and tool-using traditions are found among chimps in different parts of Africa. Even if a certain food is freely available in two areas, it may be eaten in one place and not another. It seems that those feeding, tool-using and other traditions may be passed from one generation to the next by observation, imitation, and practice. Infant chimps are intensely curious, and watch closely the foods eaten, the manner of eating them, and tools that are used for acquiring food (or any other purpose). They often then perform — or try to perform — the actions they watched. Behaviors passed on in this way may be described as primitive cultures.

Cannibalism has been recorded in chimpanzees in Tanzania and Uganda. Occasionally adult males will kill and partially eat the infant of a female of a neighboring social group. At Gombe, one mother infant pair during a 4-year period, were seen to kill and eat five newborn infants of females of their own group. Five other newborns disappeared and it is thought that they suffered the same fate. Since then, two different females have been seen hunting newborns on two different occasions: both times they failed. This behavior is not understood.

CONSERVATION ISSUES

In 1960, the eastern shore of Lake Tanganyika, some 300 miles long, comprised forested hills dropping from the Rift Escarpment to the lake. From the Burundi border in the north to the southernmost part of the Mahali National Park in the south, there were some 100 miles of chimp habitat. If one climbed to the top of the rift and looked eastward, again there were forested slopes, with just a few small villages, as far as you could see. Today, cultivated fields press up to the boundaries of the tiny 30-square-mile Gombe national park on three sides: the western boundary runs along the lake shore. Outside the park there has been almost total deforestation. With the tree cover gone, the soil has become infertile. Moreover, during the rainy season the thin layer of top soil erodes down the rocky hillsides into the valleys and the lake, where it silts up the fish breeding ground. Some places that were forested ten years ago now look like a desert, and the chimpanzees have long since gone. Only about 100 Gombe chimpanzees remain, isolated in their patch of forest, and doomed because their gene pool is not big enough to be viable.

The situation has deteriorated to this extent partly because of population growth. But this troubled part of Africa also has a terrible refugee problem.

Refugees have poured in from Burundi, 20 miles to the north, and from the Democratic Republic of Congo in the west. The people living here are beginning to face starvation because they are far too poor to purchase food from other areas where it is more suitable to grow. They cannot move as they would have done previously, because the land is already occupied unless they move south, which some of them are doing.

It is difficult to protect the precious 30 square miles of Gombe when the people around the park are starving. Conflict between humans and wild animals is destroying much of the natural habitat across Africa and other parts of the developing world. Even national parks and reserves in the developed world are not safe from the greed of those who find oil and other minerals under the surface of the supposedly sacred land.

A conservation and education program has been started that focuses on tree nurseries and agroforestry in 30 villages around Gombe. A team of Tanzanians that speak local dialects introduce conservation and education concepts. A group of women are employed to teach village women about farming methods more suitable to the terrain. Trees that give instant profit, like fruit trees and fast growing trees for firewood, are grown. Indigenous plants are reintroduced and attempts to control and prevent erosion are made.

Improving womens' self esteem through education and by raising money for the family eventually leads to decreased family size. Primary health care, especially for women and children, is also part of the program because women cannot plan a family unless they expect their children to live. Family planning and AIDS education are also included.

Many local people are employed in the park to observe the chimpanzees. They write detailed notes and use 8-mm video cameras and are proud of their work. This local pride may be why Gombe does not traditionally have poaching, while the primary threat to primates in other parts of Africa is the bush meat trade, the *commercial* hunting of wild animals for food.

CONCLUSIONS

Many chimpanzees end up in medical research laboratories because they are our closest living relatives, because their bodies are more like ours than that of any other living creature, and because they can be infected with diseases otherwise unique to us, like AIDS and hepatitis. Whether or not we think it is ethical to use them, the conditions under which most of them are maintained still need to be greatly improved. The contribution of animal nutrition to the well being of captive chimpanzees is paramount. The chimpanzee is a creature that looks out at the world around and is continually questioning, is continually fascinated. What goes on in the mind of a wild chimpanzee as he contemplates the raindrops bouncing off his hand in a storm? We will never know, but of one thing we can be sure: there is a mind and it is not that different from ours.

1

Landmark and Historic Contributions of NRC's Committee on Animal Nutrition

DUANE E. ULLREY
Michigan State University

The origins of the National Research Council's (NRC) Committee on Animal Nutrition (CAN) are embedded among the hardships of the Civil War and World War I and governmental attempts to ameliorate their effects upon national resources and the lives of Americans. The National Academy of Sciences (NAS) was mandated by Congress to enlist the brightest scientists and engineers in an honorary, non-governmental, nonprofit organization intended to serve the welfare of the United States of America and its people. Its charter was signed by President Abraham Lincoln in 1863.

Although NAS membership grew over the years to over 1,600 today, the number of persons available to offer counsel on issues of national import to government agencies or to other organizations seeking assistance was seldom adequate to the task. In 1916, a year before the United States entered World War I, President Woodrow Wilson asked the Academy to broaden its services, assist in military preparedness, and develop an organization of full-time employees to aid NAS scientists in preparing their reports for the government in a timely fashion.

Thus, the NRC was established as the working arm of the National Academies. Shortly thereafter, the NRC appointed an Agriculture Committee to deal with burgeoning needs for food and fiber. This was an exciting era in nutrition. One of the first compounds in a new class of nutrients had just been discovered and named vitamine, an amine vital for life. Later, this compound was renamed thiamin(e), and the term "vitamin" was applied to the whole nutrient class because all vitamines were not amines (Gubler, 1991). Dozens of talented specialists—among them nutritionists—were called upon to volunteer their services to the NRC, and almost immediately, the significance of animal nutrition became apparent.

THE NATION'S FIRST ANIMAL NUTRITIONISTS

In 1917, the Agriculture Committee of NRC organized a Subcommittee on Protein Metabolism in Animal Feeding chaired by Dr. Henry P. Armsby, Director of the Institute of Nutrition, Pennsylvania State College. The subcommittee produced a seven-page document entitled *Plan for Cooperative Experiments on Protein Requirements for Growth in Cattle* (National Research Council, 1917) to help resolve unanswered questions dealing with the minimum dietary nitrogen levels required to maximize productivity of cattle while minimizing usage of the supplemental proteins that were in short supply.

In 1919, the NRC Division of Biology and Agriculture formed a Committee on Food and Nutrition and divided its activities between a Subcommittee on Human Nutrition and a Subcommittee on Animal Nutrition. Dr. Armsby directed the work of this latter subcommittee until his death in 1922. Over the next four years, reports were published on experimental methods in animal production (National Research Council, 1923), the results of cooperative studies of protein requirements for growth of cattle (National Research Council, 1924), and determination of protein requirements of animals and protein concentrations in feedstuffs (National Research Council, 1926).

Following discharge of the Committee on Food and Nutrition in 1928, the Committee on Animal Nutrition was formed, and it has now been a standing committee of the NRC for over 70 years. Dr. Paul E. Howe, a nutritional biochemist with the Bureau of Animal Industry, U.S. Department of Agriculture, chaired the committee until 1941.

EARLY YEARS OF SERVICE

The early years of CAN's service coincided with the depression of the late 1920s and 1930s. Many farm families did not have electricity in their homes. There was much suffering in rural America, not for lack of electricity, but because commodity prices were so low. Farmers also were discouraged because of the long period of drought in the Central and Southern Plains — the dust bowl. Serious problems both with the welfare of our people and the welfare of our food production system were rapidly developing.

But many exciting discoveries also were taking place. During this period, riboflavin, pantothenic acid, niacin, vitamin B_6, vitamin K, cobalt, and certain fatty acids were identified as essential nutrients. And these new bits of information arrived just in time for the critical changes in animal management required by the circumstances surrounding the Second World War.

IDENTIFYING ALTERNATIVE SOURCES OF NUTRITION FOR ANIMALS AND HUMANS

As the United States was drawn into the international conflict, all sorts of items used domestically were in short supply, and many had to be reserved for military use. And because fewer people were left to serve in the productive parts of our economy, fewer people had to do more with less. Fish oil supplies that had been used to provide vitamin A in both human and animal diets were diminished. Skim milk that had been incorporated extensively into animal diets for protein, minerals, and vitamins was diverted for human use. Byproducts of meat processing, such as tankage and meat scraps, that were traditional sources of supplemental protein in animal diets often were not available. As a consequence, alternative sources of protein, minerals, and vitamins had to be identified.

Fortunately, the nutritionists and biochemists working on minerals, vitamins, and amino acids provided information vital for this purpose, and just in time. It was soon apparent that corn and soybean meal appropriately supplemented with minerals and vitamins could efficiently and economically replace the previously used more complex diets containing animal byproducts.

PROVIDING ANSWERS TO PROBLEMS OF WAR

The publications of CAN, during the 1940s, related to the problems of the time. War emergency plans for feeding cattle, poultry, and swine were devised by nutritionists who served on CAN. Because of phosphorus shortages, alternative sources were being used in animal diets, and a fluoride hazard in those alternatives was identified and quantified. In addition, the need for supplemental iodine was established as animal proteins were dropped from animal diets. Through the work of nutritionists serving as the nation's experts on CAN, the U.S. population was assured that the problem of restricted animal feed supplies was by no means insurmountable.

DEVELOPING NUTRIENT REQUIREMENTS

In the middle to latter part of the 1940s, CAN published recommended nutrient allowances for diets of swine, poultry, beef cattle, dairy cattle, sheep, and horses. These recommended nutrient allowances included subjectively established safety factors and were intended to ensure that minimal nutrient requirements would be met under any circumstances. In a number of instances, minimal requirements were unknown, and nutrient allowances were based upon successful practical experience. Feed formulators tended to add safety factors of their own, and since judgments differed in this regard, final products were appreciably different in their nutrient concentrations.

In the late 1940s, CAN concluded that the most appropriate way to express a nutrient requirement was as the minimal dietary concentration required to support normal performance of the most demanding function. Nutrient requirements were commonly established using purified diets. Unfortunately, nutrients in natural-ingredient diets are not usually as available as they are in purified diets. Thus, it was necessary, when possible, to provide some estimate of expected nutrient bioavailability in natural dietary ingredients. Since 1953, the NRC Nutrient Requirement Series presents nutrient requirements that are supported by scientific evidence or indicates that requirements are estimates.

NRC CAN publications have now been published on nutrient requirements of swine, poultry, beef cattle, dairy cattle, sheep, horses, mink and foxes, rabbits, goats, dogs, cats, laboratory animals, fish, and nonhuman primates. Included among the laboratory animals are rats, mice, gerbils, guinea pigs, hamsters, and voles. Partial requirement data are provided on nine species of fish. Nutrient requirements of nonhuman primates present a major challenge because there are over 200 species to consider.

Individual publications in the Nutrient Requirement Series on food-producing animals have been regularly updated for many years. For example, there have been 10 editions of *Nutrient Requirements of Swine* from 1944 to 1998. The length of this report has grown from 10 to 189 pages and the references from 69 to 1,524. The first edition included recommended allowances for total digestible nutrients (TDN), crude protein, calcium, phosphorus, sodium, potassium, thiamin, riboflavin, niacin, pantothenic acid, vitamin B_6, carotene, vitamin A, and vitamin D (National Research Council, 1944). Minimum requirements for these nutrients (except TDN) plus chloride, iron, copper, manganese, iodine, selenium, vitamin E, vitamin K, biotin, folacin, vitamin B_{12}, ten essential amino acids, linoleic acid, digestible energy, and metabolizable energy are included in the 10th edition (National Research Council, 1998).

DISTINCTIVE REPORTS AND MODERN TOOLS OF NUTRITION

Publications in the Nutrient Requirement Series and CAN reports on feed composition, vitamin and mineral tolerances, energy terms, nutrients and toxins in water, selenium, and chromium have had a major influence upon the productivity and health of animals on the farms and in the homes, laboratories and zoos of America. These reports provide information and guidelines that not only are used by animal nutritionists but also have been adopted by federal and state regulatory agencies. Some of the requirement reports now provide computer-based mathematical models to predict feed intake and growth rates by melding nutrient requirements with desired animal performance. These are major advances in animal nutrition, but prediction models must be based on a foundation of substantial scientific data. Mathematical equations are not a substitute for the basic information necessary to develop them, and the animal nutrition community is actively engaged in identifying and producing the needed data.

PERSONAL PERSPECTIVE

Throughout its history, the service of nutritionists on CAN has been *pro bono*. Scientists are selected by their peers, based on demonstrated professional qualifications, judgment, and ethics. They come from universities, industry, and government. In total, 117 scientists have served on the Committee, not counting the hundreds that have served on various CAN subcommittees and task forces. There are typically 10 to 15 members of CAN, serving staggered 3-year terms. Since 1941, there have been 12 CAN chairpersons.

The collective effort of the wide array of scientists and nutritionists who have been appointed to serve on CAN committees and subcommittees has contributed invaluably to the social, economic, and physical well-being of Americans throughout CAN's history. This effort also has provided information essential to the welfare, productivity, and protection of domestic and wild animal species. A former NRC staff officer, Selma Baron, may have said it best —

> "The changes that have occurred since the inception of reports on the nutrient requirements of domestic animals reflect how the lives of all Americans have been altered during the past 70 years. From the rudimentary pencil and paper calculations of essential nutrients to the use of computers to generate requirements at various stages of the life cycle, the data developed by species subcommittees of the Committee on Animal Nutrition have provided essential information to those who raise animals for food, as companion animals, and for use in research. The mechanisms for report generation may have changed, but the dedication of the animal nutritionists and other scientists has not. Their willingness to devote hours to the Committee on Animal Nutrition in preparing reports for public distribution has enriched all our lives."

REFERENCES

Gubler, C.J. 1991. Thiamin. Pp. 233-281 in Handbook of Vitamins, L.J. Machlin, (ed.) Marcel Dekker, Inc., New York.

National Research Council. 1917. Plan for Cooperative Experiments on Protein Requirements for Growth in Cattle. Washington, DC: National Academy of Sciences.

National Research Council. 1923. Experimental Methods. Washington, DC: National Academy of Sciences.

National Research Council. 1924. Growth and Recommended Nutrient Allowances for Cattle. Washington, DC: National Academy of Sciences.

National Research Council. 1944. Recommended Nutrient Allowances for Swine. Committee on Animal Nutrition. Washington, DC: National Academy of Sciences.

National Research Council. 1998. Nutrient Requirements of Swine, Tenth Revised Edition. Committee on Animal Nutrition. Washington, DC: National Academy Press.

2

Protecting Animal Health and Well-being: Nutrition and Immune Function

KIRK C. KLASING
University of California, Davis

The immune system protects animal health and contributes to animal well-being. Nutrition is an important modulator of immune function and can often tip the balance between health and disease. Current Committee on Animal Nutrition (CAN) reports provide important recommendations on nutrient requirements that account for desired or expected animal performance. Industry professionals and university scientists would like to examine other major functional endpoints. If animal health was selected as the functional endpoint instead of growth rate or reproductive performance, nutrient recommendations might change (Figure 2-1). In most cases, National Research Council (NRC) recommendations for estimated requirements are probably adequate for optimal animal health and well being, but the goal of CAN committees should be to look for exceptions.

Animal nutrition research often relies on information from human studies to point to nutrients that may require more in-depth examination. Whereas information from human studies is important (Gershwin et al., 2000), laboratory animal studies give us a better understanding of mechanistic and quantitative aspects of the immune system, such as how much of a specific nutrient is needed for the immune system to do its job and how insufficiencies or excesses of nutrients affect the immune system. It is also important to determine what

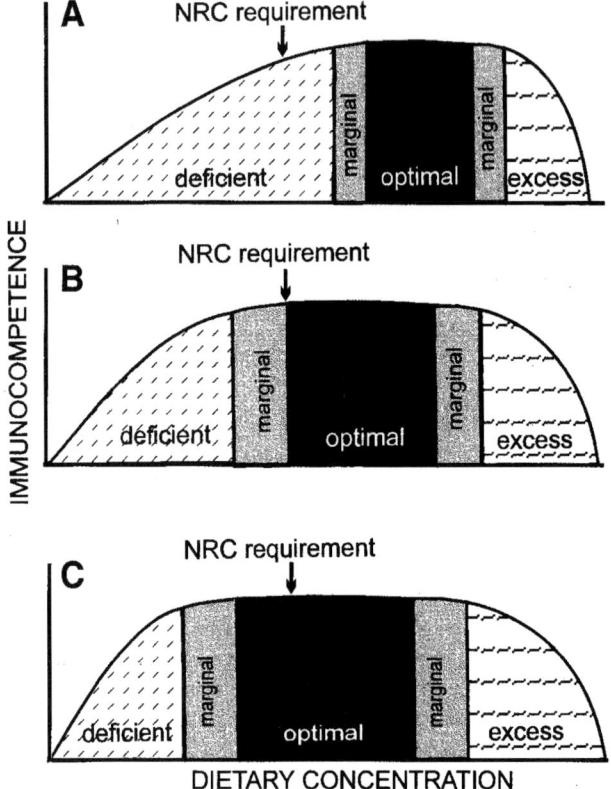

FIGURE 2-1. Dietary requirements set by the National Research Council (NRC) (shown as arrows) are usually based on concentrations that maximize growth and reproduction and prevent known deficiency pathologies. Optimal immunocompetence may occur at nutrient levels that are: higher than the NRC requirement (panel A); equal to the NRC requirement (panel B); or less than the NRC requirement (panel C).

components within the immune system should be measured and what functional endpoints are the best indicators of animal health.

Eventually target animal species need to be examined to determine specific concentrations of nutrients for optimal immunocompetence and health. Field tests and epidemiologic studies in production facilities will be necessary to verify the results with authentic disease challenges. Achieving this goal will be complicated because each species of animal has potential pathogens because and the defense to these organisms may respond differently to nutritional manipulations. Most veterinarians track 20 diseases and nutritionists are concerned with 45 or more nutrients at any given time. So, the number of combinations in one species is formidable. Sorting out these interactions between nutrition and disease can be simplified by focusing on the immune system and the modulation of this system by dietary components (Klasing, 1998).

ASSUMPTIONS REGARDING IMMUNOLOGY RESEARCH

When examining the immune system, a correlation between immunity and disease resistance is assumed. This is true for most infectious diseases, but sometimes other physiologic systems, such as membrane integrity and types of receptors on epithelial cells, are the major deciding factors.

It is also assumed that scientists know what components of the immune system to measure. This is not necessarily true. In the last 15–20 years, major advances in immunology have been driven by interest in heart disease, cancer, and basic science. These studies have given scientists a better understanding of what components to measure, but uncertainty still remains when attempting to measure animal health and susceptibility to disease. Recent efforts by toxicologists designed to find endpoints of immunocompetence that best correlate with disease resistance (Dean, 1997) should be noted by animal nutritionists as they design their research.

DIETARY INFLUENCES ON THE IMMUNE SYSTEM

Fundamental mechanisms through which diet influences the immune system and various mechanistic aspects need to be examined to determine how and why diet affects the immune system (Table 2-1). Among some of the areas to be examined are substrate needs, nutritional immunity, direct regulatory effects, hormone balance, pathology, and non-nutrient components of feeds. Examining these areas closely may provide answers to immunologic and other questions as well.

Substrate Needs

Surprising little information is available that indicates how big the immune system is, what it needs to do its job, or even its priorities for nutrient use when they become limiting. The immune system is relatively small when estimated as a percentage of body weight. The weight of all the leukocytes and their products involved in immune function are probably less than 2 percent of body weight. Although the immune system is relatively small, it is a vital system whose requirements must be met.

TABLE 2-1. Mechanisms by which nutrition modulates immunocompetence and disease resistance.

Mechanism	Nutrients
The supply of substrates to the immune system	All nutrients
Deprivation of nutrients from pathogens (nutritional immunity)	Iron, biotin, manganese
Direct regulatory effects on cells of the immune system	Fatty acids, vitamins A, D, E
Changing the balance of hormones that regulate immunity	Energy, protein, meal patterns
Reduction of pathology induced by an immune response	Antioxidants
Physical and chemical actions of feeds in the intestines	Non-starch polysaccharides, lectins, sugars

The diet supplies substrates, such as energy and amino acids, that contribute to the development, maintenance and use of the immune system. Substrates are necessary for the anabolic activity of immune system's cells (leukocytes), such as proliferation and antibody production as well as the secretion by the liver of large quantities of immunologically active molecules, the acute phase proteins. In young animals, a severe deficiency of virtually any nutrient impairs many indices of immunocompetence (Cook, 1991). Such a situation is rare in modern animal husbandry, and questions that consider the needs of the immune system relative to other systems (e.g., growth) are most relevant.

Though an understanding of how the cells of the immune system obtain their nutrients is incomplete, it is beginning to appear that the immune system uses many of the same types of glucose and amino acid transporters that nervous and other high priority tissue use. When leukocytes become activated, they express high levels of nutrient transporters, which allows them to easily obtain necessary nutrients even when they are at low concentrations. The immune system can also appropriate nutrients from muscle and other tissues. When leukocytes become stimulated by pathogens, they release a series of pro-inflammatory cytokines like interleukin 1, tumor necrosis factor and interleukin 6 that go throughout the body and redistribute nutrients—especially those from skeletal muscle. The small size of the immune system, its capacity to appropriate nutrients from other tissues, and its endowment with high priority nutrient transporters generally indicate that the immune system can usually obtain many of the nutrients that it needs to do its job over a wide range of dietary levels. However, some trace nutrients such as iron, copper, and zinc are problematic because of their low concentration in muscle and their relatively high need within the immune system. Evidence is accumulating that the dietary requirement for some trace nutrients may be higher for optimal immune function than it is for maximal growth or reproductive performance. Clearly more work is needed to determine the quantitative nutrient requirements of the immune system for its development in the young animal, its maintenance at times of good health, and its use during a challenge by a virulent pathogen.

Nutritional Immunity

Nutritional immunity is the process whereby the body withholds essential nutrients from pathogens to reduce their rate of replication. For example, it is well documented that injecting or orally feeding baby pigs iron provides additional amounts of this limiting nutrient that enhance the growth of pathogens. In this case, pathology takes the form of increased severity and duration of diarrhea. These pigs are also more likely to die. In birds, a similar situation exists within the egg to deprive bacteria of nutrients so that they are unable to colonize the albumen and infect the developing embryo. Immune cells sequester trace minerals, such as manganese and iron, when they engulf pathogens, and this action serves to starve pathogens and prevent their replication. As there are situations where high levels of specific dietary nutrients compromise immune function, the idea that "more is better" should be applied with caution.

Direct Regulatory Effects

Cellular communication within the immune system is critical because this system is one of the most complex, dynamic, and potentially destructive systems within the body. The number of communication molecules utilized by the immune system rivals that of the brain. Nutrients in the diet can directly affect the regulatory functions of leukocytes altering the type, duration, and vigor of the immune response.

For example, dietary factors such as type of fat can change the proportion of prostaglandins and other eicosanoids that are released by leukocytes to coordinate their responses to disease challenges. This response is because the type of dietary fat changes the composition of the phospholipids in the membranes of leukocytes, and these membrane fatty acids are the precursors for the synthesis of eicosanoids. Thus, the potency and specific regulatory properties of the eicosanoids released during immune responses change with the composition of dietary fat. Fish oil is high in eicosapentaenoic acid, which causes macrophages to be predisposed to release interleukins that drive T helper cells toward a Th2 type of response and less predisposed to a Th1 type of response, especially the inflammatory response (Fritsche et al., 1999; Korver and Klasing, 1997). These divergent responses are important in defense against different pathogens. It is important to note that nutrients that effect communication within the immune system "modulate" or "change" the response, accentuating some components of the response while decreasing others; the dietary manipulations do not "boost" the entire immune system. Thus, host resistance to specific pathogens shifts - with better resistance to some pathogens, but greater susceptibility to others. In the case of dietary fish oil, the prevalence of those diseases in which protection is mediated by a Th2 response is diminished, whereas the incidence of those where protection is afforded by the inflammatory response is increased.

Many other examples demonstrate how diet can affect the communication and regulatory decisions made by cells in the immune system. While lipids are one of the best demonstrated areas, there is developing literature on the regulatory actions of vitamins A, D, and E, xanthophylls, as well as some amino acids and bioactive minerals.

Hormone Balance

Diet also affects the balance of various hormones that modulate the immune system. Feeding regimes markedly affect insulin, glucagon, glucocorticoid, and IGF levels, which can change the type and duration of the immune response. For example, chronic severe feed restriction results in elevated levels of glucocorticoids, which impinge on T-cell function and decrease many indices of immunocompetence. However, very moderate restriction of food intake can increase immunocompetence and decrease the incidence of infectious diseases. When broiler chickens are restrictively fed, insulin levels are decreased and glucagon levels are increased. Changes in these hormone levels affect the chicken's ability to mobilize neutrophils, which affects their resistance to various types of disease. Other dietary factors that impact immunity through their effects on hormone levels include protein to calorie ratios and presenting food ad libitum versus in a few large daily meals.

Pathology

The immune system releases a variety of noxious substances at the site of infection in order to kill invading pathogens. Immune systems respond in a measured way so lethality is localized to the pathogen and not on surrounding tissues. However, collateral damage to healthy cells in the area surrounding the site of infection is evident in many infections. Nutritional factors that minimize the extent of pathology induced by immune responses mitigate the nutritional costs for repair and convalescence. For example, reactive oxygen intermediates released at the site of infection can cause damage to the cell membranes of healthy host cells and adequate levels of dietary antioxidants minimize this pathology (Chew, 1995).

Non-Nutrient Components of Feeds

Feeds have specific effects on immunocompetency when they contain non-nutrient substances that influence the function of leukocytes, the integrity of the intestinal epithelia, or the population of commensal microflora found in the intestines. Feed components with these activities include sugars, lectins, and other mitogens, lignins and silicas. Scientists are just beginning to understand what non-nutrient components in feeds are important and what levels of these components cause beneficial and detrimental immunomodulation.

CONCLUSIONS

The immune function of all animals should be optimized to ensure health and welfare of the individual and of the group. An optimal immune response can be measured and can occur only under appropriate conditions. Optimizing the immune system is important because responses with the wrong leukocyte populations or under-responsiveness can increase the incidence of infectious diseases, whereas over-exuberant responses result in a variety of problems in an animal. In fact, human diseases of aging such as arthritis and arteriosclerosis are caused by the immune system. Production animals can develop anorexia, impaired growth, and other systemic stress responses that reaffirm that more is not always better and what is optimal is not necessarily maximal. An effective method of assessing and maximizing the health of animals is necessary.

The ultimate goal of research efforts should be to minimize the nutrition-disease cycle, where poor nutrition causes poor immunocompetency. Poor immunocompetency, in turn, can result in greater incidence and duration of infections, which cause decreased food intake, nutrient losses, and impaired animal health and well-being.

REFERENCES

Chew, B.P. 1995. Antioxidant vitamins affect food animal immunity and health. J. Nutr. 125:1804S-1808S.

Cook, M.E. 1991. Nutrition and the Immune Response of the Domestic Fowl. Cr. Rev. Poult. Biol. 3:167-190.

Dean, J.H. 1997. Issues with introducing new immunotoxicology methods into the safety assessment of pharmaceuticals. Toxicology 119:95-101.

Fritsche, K.L., M. Byrge, and C. Feng. 1999. Dietary omega-3 polyunsaturated fatty acids from fish oil reduce interleukin-12 and interferon-gamma production in mice. Immunol. Let. 65:167-173.

Gershwin, M.E., J.B. German, and C.L. Keen. 2000. Nutrition and immunology: principles and practice. Humana Press, Totowa, New Jersey.

Klasing, K.C. 1998. Nutritional modulation of resistance to infectious diseases. Poultry Sci. 77:1119-1125.

Korver, D.R., and K.C. Klasing. 1997. Dietary fish oil alters specific and inflammatory immune responses in chicks. J. Nutr. 127:2039-2046

3

Designing Foods: Feeding Animals to Reduce Human Health Risks

BRUCE A. WATKINS
Center for Enhancing Foods to Protect Health
Purdue University

The public is concerned about how diet impacts health and risk for disease, but they are often confused when presented with conflicting reports in the news media. It is sometimes difficult for media reporters and writers to evaluate scientific and clinical studies; thus, important messages may be overlooked, misinterpreted, or down-played.

Designing foods—feeding animals to create nutritionally modified food products that improve health or reduce human health risks for disease—is an increasingly important aspect of animal nutrition. The food industry has been driven to take the lead in designing animal products because of the increasing role that foods and food ingredients play in disease prevention. Also, consumers have become more interested in self-medication. An interest in improving the quality of life of the growing aging population and reducing the amount of money spent on health care costs and treating disease have further contributed to the development of designed foods.

EARLY IMPROVEMENT IN FOODS

One of the first nutrients altered by the food industry was fat, because of its implications in cardiovascular disease, stroke, and cancer. Food scientists and human nutritionists introduced low-fat foods, which led to a decline in saturated fat intake. The composition of beef, pork, and poultry products has been altered through genetic selection—selecting animals that have less fat and are more efficient at converting nutrients to lean mass—and through the identification of nutrition requirements by the National Research Council's Committee on Animal Nutrition. Milk was improved with the introduction of low fat varieties and by fortification with vitamin D, which increased the bioavailability of calcium.

IMPACTS OF MODIFIED PROCESSED FOODS

Unfortunately these low fat foods are not decreasing the incidence of obesity, which has been on the rise in the United States. Eating less animal fat and more plant oil has increased the ratio of n-6 to n-3 polyunsaturated fatty acids in the human diet, which, when based on biochemical data, favors inflammatory responses that contribute to cardiovascular disease, some cancers, and bone disease. Processed foods and hydrogenated vegetable oils also contain less n-3 (omega-3) fatty acids and may contribute to these inflammatory processes. In support of these concerns, a direct correlation is evident when the incidence of cardiovascular disease is plotted relative to fat intake. However, the relationship between dietary fat and chronic disease indicates that both total fat and fat type influence risk for disease.

POTENTIAL DISADVANTAGES OF REDUCED INTAKE OF ANIMAL-DERIVED FOODS

Reduced consumption of animal products may actually be harmful because they supply a variety of important vitamins, high quality proteins, and minerals for growing children and elderly who have problems achieving adequate nutrition. Animal fats are an important source of arachidonic acid for children because they cannot synthesize this fatty acid, which is essential for growth and development. Docosahexaenoic and eicosapentaenoic acids, also provided in some fish oils, are important in modulating eicosanoid production and act as direct agonists of inflammatory compounds.

Current animal diets contain a limited number of different types of ingredients for simplicity and cost efficiency. Previously, poultry were fed fishmeal and other food animals were fed fish byproducts. Without these ingredients, omega-3 fatty acid concentrations decreased in animal products. Omega-3 fatty acid levels can now be restored in meat, eggs, and fish by feeding small amounts of newly developed products, such as algae-derived omega-3 fatty acids and other sources.

EXAMPLES OF CONTEMPORARY DESIGNED FOODS

Animal products that have been enriched with different types of fatty acids and fortified with vitamin E, simply by feeding sources of these nutrients to the animal, have been successfully marketed in the United States, Canada, and Australia (Table 3-1). Moreover, these designer foods are beginning to change the way people eat. One company, claiming to have one percent of the market, produces eggs that have 25 percent more vitamin E than do conventional eggs. Another company is marketing docosahexaenoic (DHA) -enriched eggs. And still another company has a line of nutrient-modified eggs.

Dairy products that have active microbial cultures, such as probiotics in yogurt and acidophilus milk, have been quite successful in Europe. Because these dairy products also make health claims, they comprise 65 percent of the functional food market in Europe (1.37 billion U.S. dollars), which is a small amount when compared with the overall food market, but a successful one in the area of designed foods.

Fatty Acids

Most recently, conjugated linoleic acid (CLA) found predominately in dairy products and meat from ruminants, has generated great interest in

TABLE 3-1. Status of designed foods.

Nutrient/Health Protectant	Target Food	Status[a]
Long-chain omega-3 fatty acids (20:5n-3, 22:6n-3)	Poultry	
	Eggs	A
	Chicken and Turkey Meat	R, D
	Pork	R
	Fish: Perch, Salmon, Trout	R
	Beef	R
	Dairy: Milk	R
Conjugated linoleic acids	Poultry: Eggs, Meat	R
	Pork	R
	Fish: Perch, Striped Bass	R
	Dairy: Milk, Cheese	R
Tocopherols, carotenoids, lycopene, phytochemicals	Poultry: Eggs	A, D
	Beef	R
	Pork	R

[a]Status: A = available, R = research, D = commercialization development

polyunsaturated fatty acid research. CLA is a derivative of linoleic acid. The two double bonds in CLA are connected in contrast to linoleic acid, which has a methylene carbon separating the double bonds in the hydrocarbon chain. CLA will not substitute for the essential fatty acid, linoleic acid, but does have potent effects on enzyme systems, eicosanoid production, and growth and development of animals that may, in the future, benefit human health. CLA seems to be important in modulating biochemical and physiologic processes in the body that may reduce the risk of cancer, heart disease, and other types of inflammatory responses with observed pathology.

Phytochemicals

Interest in creating designed foods derived from plants has risen because of the phytochemicals in plants that may reduce lipid peroxidation and protect the body from free radical damage. Phytochemicals, including carotenoids, tocopherols, phenolics, and flavonoids, are found in a variety of plants. Soybeans, garlic, cabbage, ginger, licorice, carrots, celery and flaxseed have received the greatest amount of research attention. Methods to change the composition of phytochemicals in plants, determine how they are delivered, and the nutritional quality of the final product, are being investigated.

Antioxidants

Another important area in designed foods research is in antioxidants. Antioxidants are important in lowering oxidative stress and risk of cardiovascular disease, cancers, and possibly joint disease and inflammatory arthritis. For instance, beef that has been produced with high levels of tocopherols, such as vitamin E, has increased shelf life and may contribute significantly to the health of the consumer by lowering risk of chronic disease.

FUTURE RESEARCH

Dietary fats, phytochemicals, and antioxidants have been identified as having potential properties for reducing the risk of cancer and heart disease. The real issue now is to identify how these nutrients can potentially act as modulators of tyrosine kinases, cyclooxygenase 2, signal transduction systems, oxidative stress, and transcription factors at the cellular level. Tyrosine kinases influence many activities of cells including proliferation and apoptosis. Cyclooxygenase 2, an inducible enzyme that was fully characterized in the human system 2 years ago, is responsible for the inflammatory process associated with various cancers and joint disease. Opportunities exist to use designed foods to modulate the activity of this enzyme.

CONCLUSIONS

Modifying the fatty acid composition of animal products can be accomplished. Designed foods include animal products that contain a modified nutrient or health protectant content that directly benefits human health. However, it is not clear how these health protectants impact cell activity and influence disease risk. More research is needed so that in the future health benefit labeling claims can be made and supported for the health protectants and nutraceuticals used to make these new foods.

Some of the first designed foods created could restore the omega-3 fatty acids to the human diet to balance the ratio of n-6 to n-3 fatty acids. Perhaps in the near future CLA will be better understood, and dairy products and red meat CLA levels will be increased to reduce human risk for cancer, heart disease, osteoporosis, and inflammatory diseases. Future designed foods will also involve examining raw materials and modification of nutritionally fortified processed animal products containing new health protectants.

4

Metabolic Modifiers: Advances in Economic Production of Safe Food

ROBERT J. COLLIER
University of Arizona

Metabolic modifiers are classified as compounds that alter metabolism in order to achieve an improvement in efficiency of a productive process. This may include improvements in growth rate, milk yield, body composition, or in nutrient utilization. Metabolic modifiers can have more than one function (pleiotropy), and different metabolic modifiers may cause the same biologic effect (redundancy) (Klasing et al., 1991). A challenge for animal producers is how to assess effects of metabolic modifiers on nutrient requirements of livestock.

A review of metabolic modifiers by Boyd and Bauman (1991) demonstrated that some compounds, like antibiotics, affect both growth and digestion because they influence microbial populations and digestive tract efficiency. Other compounds, such as anabolic steroids, alpha-adrenergic agonists, and somatotropins, which affect metabolism and utilization of absorbed nutrients and raise other considerations, such as effects on nutrient requirements.

These considerations are addressed in the review by Boyd and Bauman (1991), which introduced a flow chart to evaluate the impact of metabolic modifiers on nutrient requirements. The flow chart provides a key to safe, effective, and profitable use of metabolic modifiers on the farm. According to phase I (Figure 4–1), if the digestive process is altered by the metabolic modifier, then specific nutrient dietary concentrations should be altered in direct proportion to changes in digestibility. If other dimensions of animals performance are affected, then the phase II (Figure 4–2), which determines post absorptive nutrient changes, should be followed. According to phase II, if a metabolic modifier alters composition and rate of gain, then the nutrient-calorie ratio and the daily intake of specific nutrients should be altered as appropriate.

SOMATOTROPIN INCREASES MILK PRODUCTION

This flow chart (Figure 4-1) can be effectively applied to somatotropin, a 190- or 191-amino acid homeorhetic molecule that coordinates metabolism and controls nutrient flow to support a specific physiologic state (McGuire and Bauman, 1997). Somatotropin increases efficiency of growing and lactating animals through multiple sites of action. In a lactating animal, somatotropin affects a wide variety of tissues to coordinate metabolism to support increases in milk production. In a growing animal, somatotropin supports nutrient flow to lean muscle accretion and bone mass. Because lean mass is increased in growing animals, the nutrient requirements are altered. This is not the case in the lactating dairy cow where additional nutrients required are determined by the size of the increase in milk yield. There is no net change in lean body mass.

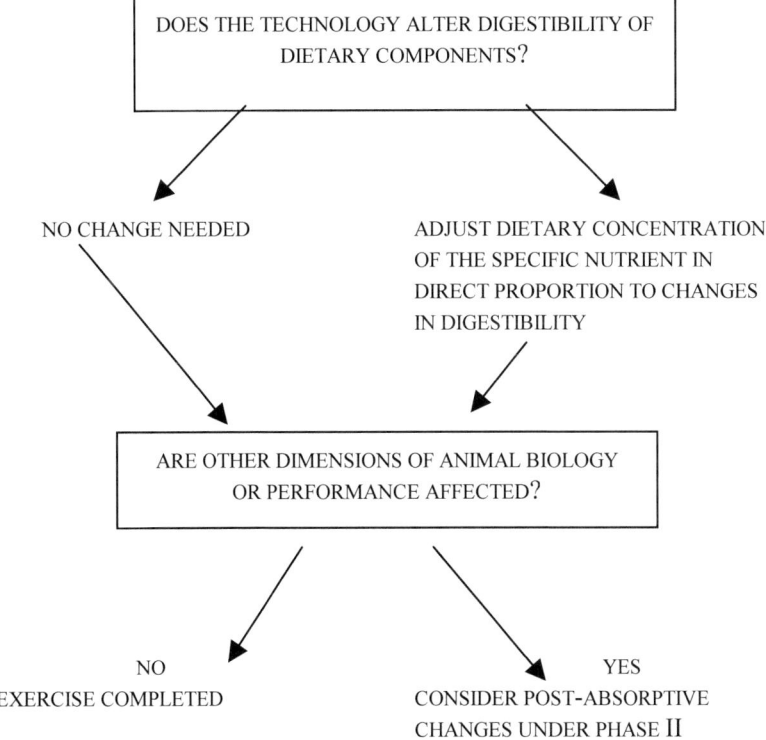

FIGURE 4–1. Phase I: Digestive Process (Boyd and Bauman, 1991).

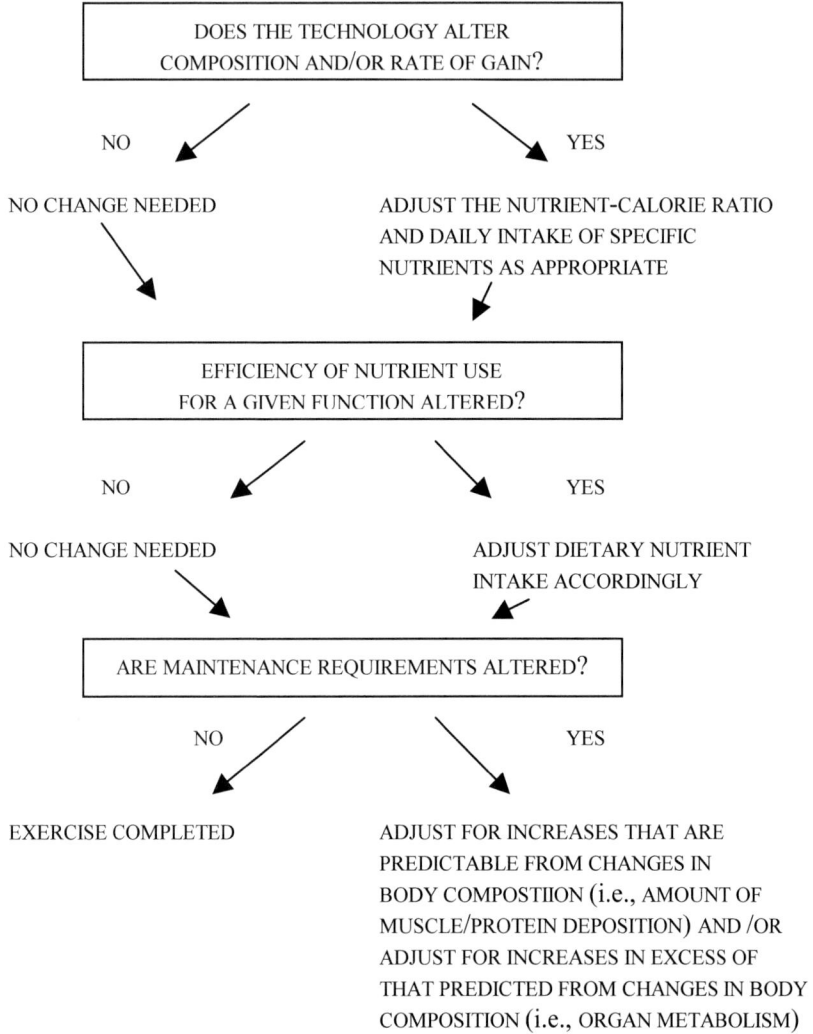

FIGURE 4–2. Phase II: Post-Absorptive Nutrient Use (Boyd and Bauman, 1991).

High-producing lactating dairy cows have higher concentrations of blood somatotropins than do lower producing dairy cows, but these levels are elevated only during lactation. This is because during lactation, the secretion rate of somatotropin from the pituitary gland is higher in high producing dairy cows (Hart et al., 1980). Injecting somatotropin in lactating dairy cows mimics rate of genetic progression. Somatotropin does not affect basal metabolic requirements or the efficiency of utilization of energy for milk production in the lactating dairy cow; so, nutrient requirements can be calculated from the increase of milk yield. Nutrients in excess of National Research Council (NRC) requirements should not produce any further benefits and research supports this claim.

Bauman and coworkers (1999) examined the impact of somatotropin use in dairy herds by matching the somatotropin customer database from Monsanto Company with the Dairy Herd Improvement Association data for the northeast region for the period of 1990 to 1998. Herds that never adopted somatotropin were compared with herds that had maintained at 50 percent or more in herd use rate of somatotropin. Herd size and milk production were relatively constant between the two groups. Herds that adopted bovine somatotropin (bST) maintained an average increase of milk production of six pounds per cow per day across the entire time period. Since only 70 percent of any of the cows could be treated at any one time (because cows are treated at 60 days into lactation) the actual increase in treated cows was actually about 10 pounds per day.

This study supported the clinical trial data. But more importantly, it demonstrated that current nutritional recommendations are adequate because dairy producers consistently maintained a milk yield response from one year to the next. Other components, such as protein were not affected -- they increased in proportion to milk yield increases. Average days in milk were similar for control and bST herds, indicating that somatotropin did not contribute to health deterioration. Decreased days in milk in 1994–1995 were attributed to feed costs and not bST.

SOMATOTROPIN ENHANCES GROWTH

When somatotropin is administered to enhance growth, maintenance requirements are increased because mean muscle mass is increased. Ideal protein intake in a set of growing boars administered porcine somatotropin (pST) was increased as lean muscle accretion increased (Etherton and Bauman, 1998).

Protein requirements also increase in ruminants but are more difficult to measure because some of the protein bypasses the rumen; some is degraded into amino acids, and some is used as microbial protein composition (Etherton and Bauman, 1998). One of several models developed to estimate ruminant protein requirements is The Cornell Net Carbohydrate and Protein System. Data for this model were developed by using actual net protein requirements as a baseline and then infusing casein into the abomasum to increase protein availability until the observed and predicted increase in actual protein were the same. While progress is being made to understand the nutrient requirements of growing animals given somatotropin, variable amino acid availability estimates have caused the response to somatotropin also be variable.

METABOLIC MODIFIERS ON THE HORIZON

Bovine Placental Lactogen

Another metabolic modifier of recent interest is bovine placental lactogen, a molecule produced during pregnancy (Collier et al., 1995). Bovine placental lactogen promotes mammary development during pregnancy. Placental lactogen binds to both the growth hormone and prolactin receptors and affects local IGF-I availability at the mammary gland. This molecule is also galactopoetic in lactating cows and promotes growth in growing animals. Growth studies utilizing both placental lactogen and growth hormone have demonstrated a synergistic effect on growth. This appeared to be mediated in part by increases in feed intake induced by placental lactogen, possibly through the prolactin receptor, which are not apparent with somatotropin alone.

Milk yield responses to placental lactogen are not as pronounced as with somatotropin, but feed intake is increased through its prolactin-like action. Because of the increase in feed intake, net energy balance is increased in lactating dairy cows treated with bovine placental lactogen. A bST-treated cow increases milk yield immediately, but feed intake does not adjust for at least 8 weeks. However, placental lactogen-treated animals immediately increase their intake. Thus, placental lactogen has potential for use early in lactation when energy balance is negative or possibly in growing animals to increase feed intake.

Conjugated Linoleic Acid

Another new metabolic modifier, conjugated linoleic acid (CLA), has real opportunity for domestic animals. The term CLA refers to a mixture of positional and geometric conjugated dienes of linoleic acid. Feeding a mixture of CLA isomers decreases fat deposition and therefore increases per cent lean mass in growing pigs (Ostrowska et al., 1999). The reduction in fat deposition reduces energy needs to achieve market weight and thus has value to producers. Interest in using CLA to improve food composition has grown since several additional properties related to human health have been shown such as its anticarcinogenic effects (Ip et al., 1994). Although Ostrowska et al. (1999) demonstrated dramatic effects of CLA on body composition, they also pointed out that improvement in feed to gain ratio was not as large as expected and may indicate an alteration in metabolic rate. Thus, much additional work is needed to estimate impact of CLA on nutrient requirements.

CONCLUSIONS

Metabolic modifiers have long been recognized as effective tools in improving efficiency of domestic animal production. However, some of these tools such as bST, antibiotics, and steroids have come under increasing pressure due to consumer health concerns. Research to identify new and improved methods of altering nutrient requirements and efficiency of production is justified and essential to continued improvements in domestic livestock production. Compounds such as CLA offer added value to livestock production by improving the quality and health benefits of consumer products.

REFERENCES

Bauman, D. E., R. W. Everett, W. H. Weiland, and R. J. Collier. 1999. Production responses to bovine somatotropin in northeast dairy herds. J. Dairy Sci. 82:2564-2573.

Boyd, R.D., D.E. Bauman, D. G. Fox, and C.G. Scanes.1991. Impact of metabolism modifiers on protein accretion and on protein and energy requirements of domestic animals J. Anim. Sci. 69(Suppl.2):56-75.

Collier, R.J., J.C. Byatt, M.F. McGrath, and P.J. Eppard.1995.Role of bovine placental lactogen in intercellular signaling during mammary growth and lactation. In: Intercellular Signalling in the Mammary Gland. Ed by C.J. Wilde et al., Plenum Press, New York, pp.13-24.

Collier, R.J., and J.C. Byatt. 1998. Somatotropin in Domestic Animals. In: Agricultural Biotechnology. Ed. A. Altman, Marcel Dekker, Inc, New York. PP. 483-497.

Etherton, T.D., and D. E. Bauman. 1998. Biology of Somatotropin in Growth and Lactation of Domestic Animals. Physiol. Rev. 78:745-759.

Hart, I.C., J.A. Bines, and S. V. Morant. 1980. The secretion and metabolic clearance rates of growth hormone, insulin and prolactin in high- and low-yielding cattle at four stages of lactation. Life Sci. 27:1840-1847.

Ip, C. M. Singh, H.J. Thompson, and J.A. Scimeca. 1994. Conjugated linoleic acid suppresses mammary carcinogenesis and proliferative activity of the mammary gland in the rat. Cancer Res. 54:1212-1215.

Klasing, K.C., W.C. Wagner, and K.W. Kelley. 1991. Impact of metabolic modifiers on target animal health and environmental safety with emphasis on somatotropin. J. Anim. Sci 69(Suppl. 2):88-99.

McGuire, M.A., and D. E. Bauman. 1997. Regulation of nutrient use by bovine somatotropin: the key to animal performance and well-being. In: IXth International Conference on Production diseases in Farm Animals 1995. ed. By H. Martens.-Stuttgart:Enke

Ostrowska, E., M. Muralithran, R.F. Cross, D. E Bauman, and F. R. Dunshea 1999. Dietary conjugated linoleic acids increase lean tissue and decrease fat deposition in growing pigs. J. Nutr. 129:2037-2042.

5

Nutrients as Regulators of Gene Expression

DONALD B. JUMP
Michigan State University

Macronutrients (carbohydrate, lipids, and protein) play a fundamental role in mammalian growth and development by serving as a source of energy as well as components for the synthesis of structural and regulatory components of cells. Certain macronutrients also affect cell function through changing circulating hormones (e.g., glucose regulation of insulin release from pancreatic β-cells). We have known for many years that certain micronutrients, like vitamins A and D, have dramatic effects on gene expression through the regulation of intracellular receptors that bind promoters of specific genes.

The notion that macronutrients, or their metabolites, might also affect gene expression is a new concept that has emerged over the last decade. It is now clear that certain macronutrients (or their metabolites) affect gene expression and lead to changes in the abundance of key proteins that function at critical steps in metabolic pathways or control cell division or differentiation (Figure 5-1). Here, I briefly describe how three macronutrients (i.e., cholesterol, glucose, and dietary fat) affect gene expression to change cell metabolism.

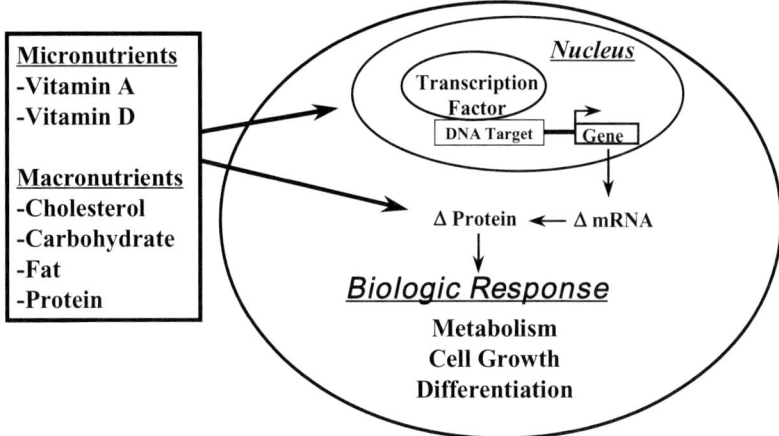

FIGURE 5–1.   Overview of nutrients as regulators of gene expression.

CHOLESTEROL REGULATION

Pioneering work of Brown and Goldstein (1997) over the last 10 years has clearly shown that cholesterol regulates its own metabolism. Cholesterol plays an important role in growth because it is a component of cell membranes and steroids. The cell has the capacity to regulate its uptake of cholesterol through the low density lipoprotein (LDL) receptor (LDLR), as well as to regulate *de novo* synthesis of cholesterol. (Figure 5–2). Low intracellular cholesterol levels prompt a cellular response that leads to an induction in both the LDLR and the enzymes responsible for cholesterol synthesis. As intracellular cholesterol levels increase, this process is reversed. These changes in cholesterol metabolism are due, at least in part, to the effect of cholesterol on the nuclear content of a family of specific transcription factors called sterol regulatory element binding proteins (SREBP). Three SREBP subtypes have been described (i.e., SREBP1a, SREBP1c, and SREBP2). SREBPs are helix-loop-helix transcription factors that bind at specific cis-regulatory elements called sterol regulatory elements (SRE) in the promoters of several genes involved cholesterol synthesis, cholesterol uptake, and fatty acid synthesis. Binding of SREBP to SRE induces transcription of specific genes, leading to an increase in the mRNA and corresponding protein.

SREBPs are synthesized as ~125 kd precursor proteins (pSREBP) tethered to the endoplasmic reticulum and golgi membranes. Specific proteases digest the precursor to generate a 65 kd form of SREBP (nSREBP) that travels to the nucleus where it binds SREs. Cholesterol regulates the nuclear content of

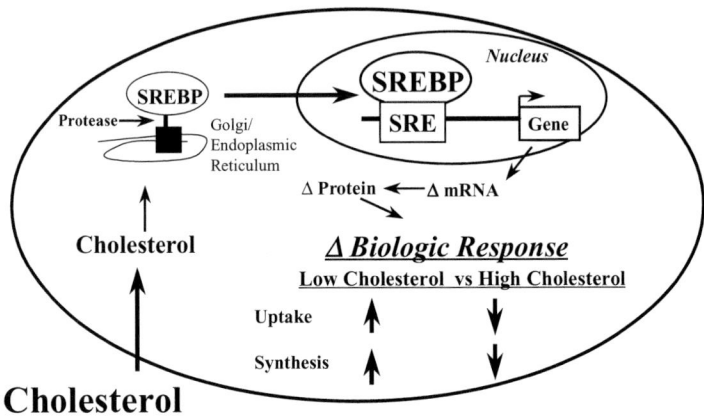

FIGURE 5–2. Cholesterol regulation of cholesterol metabolism.

SREBP by controlling the proteolysis step. Cholesterol controls the passage of SREBPs from the endoplasmic reticulum to the golgi. When cholesterol levels are low, proteolysis of pSREBP occurs and generates nSREBP and specific SREBP-regulated genes are activated. As intracellular cholesterol levels rise, SREBP proteolysis declines leading to a fall in the nSREBP and suppression of gene transcription. Thus, cholesterol is a feedback regulator for its own synthesis by controlling the nuclear content of SREBPs.

GLUCOSE REGULATION

Glucose effects on gene expression have traditionally been ascribed to its regulation of insulin release from the pancreatic β-cell. However, studies with primary hepatocytes and pancreatic β-cells have shown that glucose acts independently of insulin to control gene expression. When glucose concentrations increased in liver cells, enzymes like pyruvate kinase and acetyl-CoA carboxylase increase. This leads to increased flux of glucose metabolites into fatty acid synthesis. In the pancreatic β-cell, transcription of pyruvate kinase and the insulin gene is induced by glucose. In both cases, glucose stimulates transcription of specific genes that augment the cellular content of the corresponding mRNA and protein. These glucose-regulated genes contain in their promoters specific cis-regulatory elements, called carbohydrate (or glucose) response elements. Unfortunately, the identity of the transcription factors binding these elements and the mechanism for glucose regulation of these factors is unknown. Nevertheless, the overall effect of this regulatory mechanism is to promote glucose uptake from the circulation, its intracellular metabolism and assimilation into lipid.

FATTY ACID REGULATION

Dietary fat and the resulting fatty acids have emerged as major regulators of gene expression through at least three distinct mechanisms: 1) as precursors to eicosanoids, 2) as ligands for nuclear receptors, and 3) as a controller of the nuclear content of SREBP1c.

Eicosanoids are oxidative products of arachidonic acid, a polyunsaturated fatty acid (PUFA). These products include prostaglandins, leukotreines, and thromboxanes and involve two enzymatic pathways, the cyclooxygenase (COX) and lipoxygenase pathways. Arachidonic acid is released from cell membrane phospholipids by the action of phospholipase A2 and is converted to eicosanoids by COX or lipoxygenases. These bioactive lipids, like prostaglandin E_2 (PGE_2), are secreted from cells where they act locally on plasma membrane-associated G-protein linked receptors (GPR) on target cells (Figure 5-3). These receptors control intracellular second messenger levels, like cAMP and free calcium, which, in turn, control numerous cellular processes through changes in protein phosphorylation. Consequently, eicosanoid binding to GPR rapidly stimulates protein phosphorylation, leading to changes in metabolism, cytokine production, and production of adhesion molecules. Some of these effects involve changes in gene expression through controlling the activity of specific transcription factors, like cFos, cJun, NFκB, and cMyc.

Essential fatty acid deficiency is associated with a decline in arachidonic acid phospholipid content and the production of eicosanoids. Eicosanoid production is associated with inflammatory responses and host defense. Interestingly, certain dietary fats, particularly the highly unsaturated n-3 fatty acid, are poor substrates for COX. This leads to a decline in eicosanoid production as well as a diminished inflammatory response.

A second route for fatty acids to affect gene expression is through the regulation of a family of nuclear receptors called peroxisome proliferator activated receptors, (PPAR). Four PPAR subtypes have been identified (i.e., α, β, γ1, and γ2). These are members of the steroid superfamily of nuclear receptors that bind DNA motifs, called peroxisome proliferator regulatory element (PPRE). PPARs bind PPRE in association with a second receptor called retinoid X receptor (RXR). PPARs were first identified as the molecular targets for peroxisome proliferators. Peroxisomes are subcellular organelles involved in β-oxidation of fatty acids and cholesterol metabolism. Peroxisome proliferators are

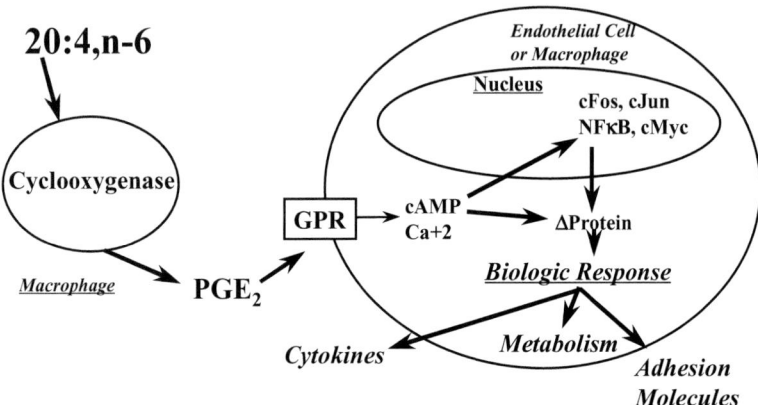

FIGURE 5-3. Eicosanoid regulation of gene expression.

a structurally diverse class of hydrophobic compounds that lead to peroxisomal proliferation in rodent liver. However, certain drugs have as their molecular targets specific PPARs. For example, the hypolipemic drugs may target PPARα, while PPARγ is the molecular target for the insulin sensitizing drugs.

PPARs have been associated with the regulation of expression of genes involved in nearly all facets of fatty acid metabolism (i.e., fatty acid uptake, fatty acid binding, fatty acid oxidation, and adipocyte differentiation). In addition, these receptors are reported to participate in inflammation as well as cell growth and differentiation. Interestingly, fatty acids, their metabolites and certain drugs bind to and activate PPARs, much like a steroid binds to a steroid receptor. For example, in the rodent liver, hypolipemic drugs and highly unsaturated n-3 fatty acids activate PPARα, leading to enhanced peroxisomal and microsomal fatty acid oxidation. In preadipocytes, thiazolidinediones and possibly eicosanoids bind to and activate PPARγ2. This accelerates the rate of adipocyte differentiation and increases insulin sensitivity of the adipose depot.

While PPARs have attracted considerable attention as molecular targets for fatty acid regulation of gene expression, it appears that these factors are not the sole targets for fatty acid effects on the genome. Recent studies indicate that one of the SREBPs, namely SREBP1c, is particularly sensitive to fatty acid regulation. Several reports appearing in the last 2 years have suggested that SREBP1c plays a major role in both hepatic and adipocyte lipogenesis, i.e., the synthesis of fatty acids and triglycerides. Feeding animals diets supplemented with polyunsaturated fatty acids suppress the mRNA encoding SREBP1c as well as both the precursor and nuclear forms of SREBP1c. Because SREBP1c is a key factor in the transcription of several lipogenic genes, its decline leads to a reduction in lipogenic gene expression and *de novo* lipogenesis.

Clearly, fatty acid effects on cell function go far beyond serving as sources of energy and structural components of membranes. Fatty acids enter cells, undergo metabolism, and can serve as ligands for both membrane and nuclear receptors. Alternatively, fatty acids or their metabolites can regulate the nuclear abundance of SREBP1c, a key transcription factor in the synthesis of fatty acids and triacylglycerols.

SUMMARY

I have highlighted some of the recent advances in macronutrient regulation of gene expression, and I have provided the detail needed to understand the roles and effects of these nutrients. In addition to their role as an energy source, as structure elements or precursors to signaling molecules, macronutrients clearly have profound effects on gene expression. This nutrient-genome interaction interfaces with other signaling networks to allow integration of cellular control between dietary intake and internal regulatory mechanisms. It reflects an adaptive response, allowing cells to adjust to changes in the type, quantity, and duration of nutrients ingested for efficient growth.

While pharmacologic agents have been developed to control cholesterol synthesis (i.e., statins) and lipid synthesis (i.e., fibrates or thiazolidinediones), a better understanding of these regulatory processes will allow for the design of more effective agents to modify metabolism in both man and animals. This understanding and ability to design agents to modify metabolism will have benefit to human health, as well as animal production and health.

REFERENCES

Brown, M.S., and J.L. Goldstein. 1997. The SREBP pathway: Regulation of cholesterol metabolism by proteolysis of a membrane-bound transcription factor. Cell 89: 331-340.

Jump, D.B., and S.D. Clarke. 1999. Regulation of gene expression by dietary fat. Annu. Rev. Nutr. 19: 63-90.

6

Our Changing Environment: Developing Strategies for the Future

DANNY G. FOX
Cornell University

North Americans have become increasingly concerned about water and air quality. In an attempt to address this problem, the U.S. Department of Agriculture (USDA) and the Environmental Protection Agency (EPA) have released a strategy for comprehensive nutrient management planning on farms for protecting water quality. This strategy sets the direction for how nutrients will be managed and how water quality is protected. This program proposes that every livestock farm have a nutrient management plan within the next 10 years. Large farms with over 1,000 animal units are being required to develop plans now. Records that document how nutrients are managed will be required. A monitoring program will also be instituted to verify that federal and state standards are met.

The amount of nitrogen and phosphorus used per unit of milk or meat sold must be decreased. Amount of nitrogen and phosphorus applied per acre will be critical because, if the nutrients are not recycled, they will accumulate and become a potential risk for leakage into ground or surface water. Therefore, viable alternatives for moving excess manure off the farm, and alternatives for processing it for that purpose are being developed and evaluated.

The USDA/EPA proposal focuses on crop and manure nutrient management more than animal feeding. I agree with this approach to address the immediate problem of keeping the manure nutrients from leaking into the surface and ground water. However, we have found there are great opportunities for reducing excess nutrients on the farm while reducing feed costs on the animal feeding side. In a case study on one farm, instituting a crop nutrient plan that focuses on efficient use of manure nutrients increased profits by $3,000. Use of a computer model, the Cornell Net Carbohydrate and Protein System (CNCPS), to accurately predict animal requirements and feed nutrients available to meet requirements while improving feeding management resulted in an additional annual increase in returns over feed costs of $42,000, while reducing nitrogen and phosphorus excretion by 25 to 33 percent.

Equations and coefficients in CNCPS and similar computer models are derived from the Committee of Animal Nutrition (CAN) reports and supplements, as well as published data. So, current summaries about nutrient requirements are important for modelers to accurately predict independent effects of variables that influence requirements and feed nutrient availability and utilization.

We have conducted studies on a range of dairy and beef cattle feeding farms (the smallest was a 40-cow pasture-based dairy and the largest were a 500-cow dairy and a 1000-head beef feedlot) to determine if there were problems with excess nutrients. Measurements revealed that 59 to 85 percent of the nitrogen from nitrogen fixation and the phosphorus and potassium from feeds was retained on the farm and not exported as milk or meat, regardless of type or size. Volatilization of nitrogen contributed to a 67 to 75 percent loss of excess nitrogen. On one farm, a leaching model, based on soil type characteristics and rainfall, predicted that 10 percent of excess nitrogen enters ground water. This was validated when we measured levels of nitrogen and phosphorus above federal water quality guidelines during the growing season in a stream containing only surface and groundwater from that farm's cropland.

Our conclusion is yes; we in agriculture have a problem with excess nutrients and we must assume ownership of its potential for impacting water quality. It will be of increasing concern, because everyone desires clean water and air. But producers must recognize that they may have problems with excess nutrients. A survey of 25 dairy farmers in 1995 revealed they believed that farms did not create environmental problems; instead, they believed the problems were created by people who talked about the environment. They also stated that they planned to continue to make decisions based on economics, and not the environment, until regulated to do so. In our state, we are working to create an awareness that there is a problem and to encourage farmers to become proactive to address the problem voluntarily.

A CASE STUDY

To better understand the problem of excess nutrients and their impact on water quality, we conducted a study at Cornell's Animal Science Teaching and Research Farm. Our data revealed a 50 percent increase in milk production with no increase in cow numbers over the 15-year period studied and with no change in crop acres or yields. As a result, this increase in milk production was supported by increased importation of feed nutrients on the farm, which increased soil phosphorus from 7 to 30 pounds per acre. One farm that applied the same amount of nutrients to the cropland as the Cornell farm did (170 lb. of nitrogen and 32 lb. of phosphorus per acre), had stream levels of 14.4 ppm of nitrogen and 0.41 ppm of phosphorus, which are above EPA standards. Four wells in corn fields sampled over the 15 years had a 54 percent increase in nitrate content. Some of the wells beginning with nitrogen levels at about half of the EPA standard now exceed it. Nitrogen in wells in the unfarmed hillsides remained low (less than 0.6 ppm, compared to the EPA standard of 10 ppm). Obviously nutrients on most farms are diluted by water from other sources and from hillsides, so they will not reach these levels. However, this study demonstrates that livestock farms do have potential to impact water quality and that producers should be concerned.

THE IMPACT OF WHOLE FARM NUTRIENT MANAGEMENT

We have been conducting studies on dairy and beef farms in the state of New York to develop a process for whole farm nutrient management planning to address the problem of excess nutrients. The first step was to develop computer software for herd, manure, crop, and soil nutrient management planning. We call this family of computer programs the Cornell University Nutrient Management Planning System (CUNMPS). The CUNMPS integrates research and experience about livestock nutrition, crop requirements, and manure management. In four dairy farm case studies, use of the herd nutrition computer program along with improved feeding management indicated nitrogen and phosphorus in manure can be reduced by up to one third, while feed costs can be reduced by $50 to $130 per cow annually. The manure, crops, and soils computer program determines the amount of manure nutrients that can be recycled by the crops, and where and when to apply them to protect water quality. Each of these programs is being used separately by herd feeding advisors and crop advisors.

Several whole farm plans have been developed with the complete CUNMPS on case study farms and are being evaluated as they are implemented. On one 500-cow dairy, intensively managed grass was substituted for corn and alfalfa on wet, erodable hillsides, with corn and alfalfa grown in the flat valley land. The grass provides a sink for excess manure nitrogen, while reducing nutrient runoff and soil erosion. In the new plan, annual cost of milk production is predicted to decrease by $40,000. With changes in herd and crop nutrient planning, the percentage of nitrogen and phosphorus feed nutrients that are purchased are expected to decline 44 and 48 percent, respectively.

On a 350-cow dairy, changing four management practices (grouping and feeding cows by amount of milk production, improving forage quality, and maximizing crop yield to the potential for the soil resource and climate) is expected to decrease the excess nitrogen and phosphorus on the farm by 44 and 51 percent, respectively, while reducing annual feed costs by over $96,000.

Our studies show that increasing milk income to overcome rising costs by using improved feeding, genetics, and management technologies to increase milk production per cow versus increasing cow numbers is better for both the environment and farm profits. In the 350-cow dairy case study, increasing milk production 10 percent by increasing milk production per cow versus increasing cow numbers was projected by the CNCPS to result in 10 percent less nitrogen and 12 percent less phosphorus, with a $34,000 reduction in annual feed cost.

IMPROVING FARM SUSTAINABILITY

Improving farm sustainability involves improving economic viability on each farm while protecting the environment. These improvements require feeding recommendations that are based on predicting animal requirements and nutrients available from feeds accurately in each unique production setting. This requires the ability to use farm specific inputs to account for variables such as animal type and level of production, feed composition, and environmental conditions. Farm plans will need to integrate all the components of that particular farm, such as animals, soils, crops, manure management, and farm business records. Complex computer models can be used for this task, given the availability of powerful computers at low cost and increased availability of information on farms such as feed intake and feed analyses, crop yields, soil tests, and information from the satellite-based Global Information System (GIS). In many situations, we now have more information than we are capable of translating into recommendations for feeding and cropping. So, CAN reports will become even more important, as there is increased need for recommendations in the form of equations and coefficients developed from published studies, which can be converted into prediction equations for use in computer models that allow for farm specific feeding recommendations.

Our experience indicates whole farm nutrient managment plans developed must be farm-specific and cost-effective, or they will not be implemented. The animal nutritionists and crop specialists now advising farmers must work more as a team to create an integrated whole-farm nutrient management plan. To provide the science-based tools needed, scientists at academic institutions must work more as a team so that the scientific information can be integrated for application rather than presented to the farmer as a series of independent best management practices. This collaboration poses a challenge to our land grant institutions because our reward system favors individual accomplishment more than team contributions and the development of new science over the integration and application of accumulated knowledge.

DEVELOPING ALLIANCES BETWEEN CONSUMERS AND FARMERS TO PROTECT WATER QUALITY

The time to address the problem of excess nutrients on farms is now, not sometime in the future. An example is the New York City (NYC) watershed. The city of New York obtains its water supply from one of the largest surface storage and supply systems in the world: 19 reservoirs and 3 controlled lakes in 8 counties north and northwest of NYC. This system provides on average 1.34 billion gallons of high-quality unfiltered water daily to the over 9 million people living in the NYC metropolitan area through a 6000-mile grid of water mains. Recently, the EPA mandated that the city build a water filtration plant that would filter all of this water, which would cost over $6 billion to build, or take steps to protect the water quality in these reservoirs, which store over 550 billion gallons of water. A Watershed Agricultural Council (WAC) was formed to develop a partnership with farms in the NYC watershed area to protect water quality. Partnerships were developed with Cornell Cooperative Extension and University Scientists, Soil and Water Conservation Districts, and the USDA Natural Resources Conservation Service to provide technical support and tools for developing whole farm nutrient management plans. The goal of these plans is to minimize nutrients, pathogens, and sediments leaving the farm in surface and ground water. The NYC Department of Environmental Protection provides funds for developing whole farm plans to meet water quality goals, scientific support for managing on farm pollution sources, and implementation of all structural changes and management of the plan as approved by the WAC. Participation is voluntary. However, because of the commitment of NYC to share the cost, a large number of farms in the NYC Watershed now have nutrient management plans in place. Similar programs have been and are being developed in other watersheds in the state.

CONCLUSIONS

Nutrient management planning can be a win-win for agriculture, the environment, and the U.S. population. A key component of protecting water quality is reducing nutrient loading through more accurate prediction of nutrient requirements. CAN's continual independent and objective advice is increasingly important for development of computer models that provide the biologic basis for predicting farm specific nutrient requirements, which allows economics and excretion rates with alternatives to be accurately predicted. When combined with whole farm planning, nutrient losses into the environment can be reduced.

REFERENCES

Fox, D.G., T.P. Tylutki, M.E. Van Amburgh, L.E. Chase, A.N. Pell, T.R. Overton, L.O. Tedeschi, C.N. Rasmussen, and V.M. Durbal. 2000. The Net Carbohydrate and Protein System for evaluating herd nutrition and nutrient excretion. Animal Science Mimeo 213. Department of Animal Science, Cornell University, 130 Morrison Hall, Ithaca, New York 14853-4801.

Hutson, J.L., R.E. Pitt, R.K. Koelsch, and R.J. Wagnet. 1998. Improving dairy farm sustainability. II. Environmental losses and nutrient flows. J. Prod. Agric. 11:233-239.

Klausner, S.D., S.D., D.G. Fox, C.N. Rasumssen, R.E. Pitt, T.P. Tylutki, P.E. Wright, L.E. Chase, and W.C. Stone. 1998. Improving Dairy Farm Sustainability. I. An approach to animal and crop nutrient mangement planning. J. Prod. Agric. 11:225-232.

Tylutki, T.P., and D.G. Fox. 2000. Managing the dairy feeding system to minimize manure nutrients. Northeast Agricultural Engineering Service Bulletin 130. In: Managing Nutrient and Pathogens from Animal Agriculture. NRAES, 152 Riley-Robb Hall, Cornell University, Ithaca, NY 14853.

Wang, S.J., D.G. Fox, D.J.R. Cherney, S.D. Klauser, and D.R. Bouldin. 1999. Impact of dairy farming on well water nitrate level and soil content of phosphorus and potassium. J. Dairy Sci. 82:2164-2169.

Wang, S.J., D.G. Fox, D.J. R. Cherney, L.E. Chase, and L.O. Tedeschi. 2000. Whole herd optimization with the Cornell Net Carbohydrate and Protein Sytem. III. Application of an optimization model to evaluate alternatives to reduce nitrogen and phosphorus mass balance. J. Dairy Sci. 83:2160-2169.

7

Readiness of Military Service Animals

SUSAN YANOFF and MICHELLE ROSS
U.S. Department of Defense

MILITARY WORKING DOGS

The Army, Navy, Airforce, and Marine Corp have 2,000 military dogs stationed worldwide. These dogs are used to perform patrols and to detect explosives and drugs, among other duties. The dogs most commonly used for these activities include Belgian Malinois (herding dogs similar to German shepherds), German shepherds, and Dutch shepherds. Retrievers are also used as working animals by the Department of Defense (DoD). These dogs, especially the Malinois and shepherds, possess characteristics that make them well-suited for military service: size, intelligence, and adaptability. They are also an aggressive breed that can be trained to protect by utilizing their special senses and controlling their catching, biting, and holding behavior. Dogs that meet height and weight requirements are often obtained from farms around the Czech Republic and Hungary countryside.

Potential military working dogs undergo extensive evaluation prior to procurement by the DoD. If they are deemed acceptable for potential service, they are admitted to the Training Squadron at the DoD's Official Military Working Dog Training School in Lackland Air Force Base, Texas.

While military working dogs might be considered "weapons," they are not autonomous; they have a working interdependence with their handler. The dogs are stationed with soldiers worldwide and may also be deployed for a period of 1 to 6 months.

Military working dogs must adapt to extreme temperature changes and often rugged, unyielding environments. While military working dogs may not exert as much energy per day as sled dogs or active hunting dogs, they are required to perform critical and tedious duties.

Nutritional Factors

A high-quality diet helps military working dogs remain healthy as they deal with their environmental and physical challenges. A nutrient-dense, highly digestible, high-performance diet is provided by a commercial supplier for the daily rations of military working dogs.

The care and feeding of military working dogs present situations that are much different from those associated with the care and feeding of pet canines. Unlike our pets, but just as military soldiers do, dogs must carry their food with them when they deploy. Feeding a nutrient-dense food is convenient because it keeps the amount of food to be transported to a minimum. Highly digestible food also creates less waste, which is a consideration as military quarters can be limited.

The diet provided to military working dogs is a fixed formula. This helps to ensure consistency, which is important for preventing gastrointestinal problems encountered by alterations in quantity and quality of nutrients fed. The number one emergency and a common cause of death in military dogs is gastric dilation. During gastric dilation, the stomach fills up with air and twists on itself. If this condition is not treated immediately, the dog quickly dies.

Just as humans in the military fall ill, military working dogs can become sick. When this happens, special diets are provided that meet the dog's special nutrient requirements until the dog recovers. If the health condition is a chronic one, such as a kidney problem that would require specialized long-term care, the dog is retired from service.

Monitoring the health of the military working dogs is important to ensure their military readiness. Dogs are monitored and maintained at ideal weight. Should the animal gain or lose more than 5 pounds, it is examined for medical problems. If none are found, the diet is immediately adjusted and caloric intake is established to return the dog to its optimal weight.

Many do not appreciate that nutrition is a major factor that must be considered in the health and welfare of military service dogs. The form and composition of the diet must be considered to allow for rapid deployment. The diet must also promote health, reduce risk for disease, maintain immune function ,and ensure that the dogs are physically fit to serve in the military.

MILITARY MARINE MAMMALS

The U.S. Navy has a large collection of marine mammals that are used as military working dogs of the ocean. The Navy also studies the diving physiology, hydrodynamics, and sonar capabilities of marine mammals. Whale and dolphin sonar is studied because nothing can be developed electronically that could mimic these living systems.

Nutritional Bioenergetics

Bioenergetics, the study of energy transformations in living systems, connects physical sciences, such as thermodynamics, with biological energy flow. Nutritional bioenergetics explores the relationship between dietary components and energy expenditure. Energy balance is predicted based on body weight, gender, activity, environment, and food nutritive value.

The objective of nutritional bioenergetics for performance animals is to optimize maintenance, growth, reproduction, and athletics. However, there is very little published information about nutrient requirements of marine mammals, which makes this objective difficult to achieve. Although the National Research Council (NRC) has published a number of reports on multiple animal species, that are a great source of information, there is not an official NRC source on marine mammal nutrition.

Most captive marine mammals utilize feeding programs based on a defined fish weight from which kilocalories can be calculated. However, this method is not optimal because fish type and season of the year cause tremendous variation in the caloric content of any given fish species.

Marine mammals have high-energy requirements, which are 1.5 to 2 times greater than terrestrial animals of similar size, in part because of thermoregulatory requirements. Thermoregulation is the mechanism that warm-blooded animals evolved to maintain their narrow body temperature range when faced with different environmental conditions. Living in a water environment presents a unique challenge because of water's high heat capacity, which draws heat away from the body 25 times faster in water than in air.

Dolphins, whales, and some other marine mammals have developed physiologic adaptations to compensate for this challenge. A cross section of skin and tissue in the dolphin reveals a small layer of subcutaneous fat and a much larger layer of hypodermis or blubber. Blubber is fibrous connective tissue imbedded with fat that acts like a wet suit and insulates against the cold. It is also a nutritional source, is highly labile, and can expand and contract rapidly depending on the environmental challenge -- the depth of the layer can change markedly within a couple of weeks.

Lactation also increases energy requirements. Dolphins don't drink water but derive all their water from their food -- fish are generally 75 or 80 percent water. Metabolic water is also produced during digestion and metabolism.

Determining Nutrient Requirements

Few controlled studies have been conducted to determine nutrient requirements of sea mammals; so, military researchers have looked at basic nutritional challenges. Specifically, the effects of high and low fat diets were compared with weight, body condition, and metabolism. Dolphin populations were divided into groups based on gender, age, and activity.

Older mammals require fewer calories for maintenance than do younger animals because the body becomes fatter with age. Animals fed high fat diets had higher body fat than did animals fed the same amount of calories from other sources. Metabolic rate decreases with age, but low fat diets increase metabolic rate.

Free ranging aquatic animals cannot be put in a metabolic chamber to measure energy expenditure; so, stable isotopes of water, deuterium, and ^{18}O must be administered orally. This method is effective but expensive—an average dose costs $6,000. This technique allows very accurate measurements of energy expenditure, and metabolic rate can be calculated in free ranging animals.

More research needs to be conducted on the nutrient requirements of marine mammals. Dolphins are specialized mammals and fill a unique military service and ecological niche. Early evidence suggests that bioenergetic principles in marine mammals mimic the bioenergetic principles in terrestrial mammals, supporting the concept of evolution and underscoring the relationship among all living animals.

CONCLUSIONS

The use of animals in the military has been a significant and critical component of our nation's protection and defense throughout history. These animals serve in specialized capacities, working closely with their human counterparts. Defining and meeting their nutrient needs continues to be of the utmost importance to ensure that the non-human members of our military forces are well cared-for and given the necessary tools for optimum performance.

8

Research and Education Needs for the Next Generation

QUINTON ROGERS
University of California, Davis

There are many tools that are used in research and teaching at our universities. One of those tools, the National Research Council's (NRC) nutrient requirement series, represents the primary publications of the Committee on Animal Nutrition (CAN). These publications have been used throughout most of this century for research and education purposes. The continual update of the reports in this series is critical to our next generation of scientists and educators.

PAST GENERATIONS

The number of publications by CAN that have been used by past generations of researchers and educators peaked in the mid 70s (Figure 8-1), and since that time the interval between revisions of the publications has increased and the number of publications produced per year has decreased. These trends reflect the fact that the rapid pace of science produces more material that must be reviewed with each revision, which requires more time and resources. For example, the number of pages (Figure 8-2) and number of references (Figure 8-3) in each publication has increased steadily throughout the 70 years that CAN has existed. The improved technology of communication has resulted in revisions of the reports on food-producing animals that have evolved from static documents containing tables with numerical values to become more dynamic with the incorporation of computer models. This should not only make the reports more useful but also should extend the useful life of the report.

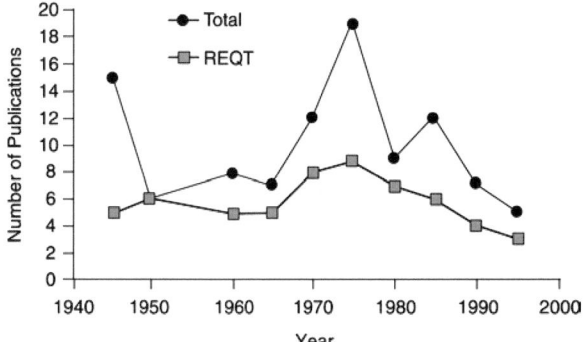

FIGURE 8-1. Total number of CAN publications and those in the Nutrient Requirement Series per 5-year period.

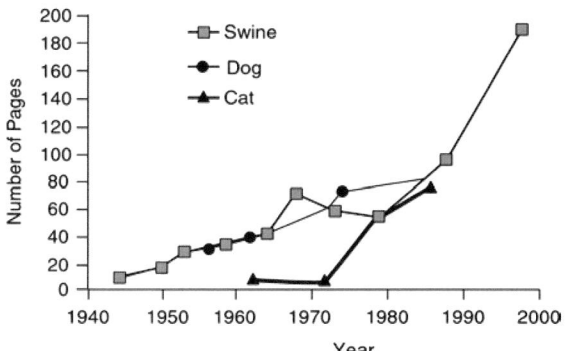

FIGURE 8-2. Number of pages in the National Research Council Nutrient Requirement Series for reports on swine, dogs, and cats during the last 45 years.

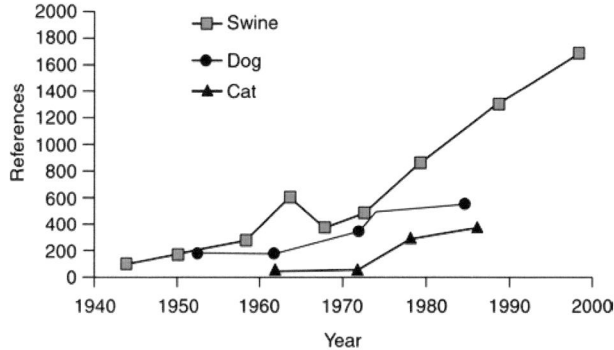

FIGURE 8-3. Number of references in the National Research Council Nutrient Requirement Series for reports on swine, dogs, and cats during the last 45 years.

It should also be noted that the reports are produced by experts who volunteer their time because they care deeply about the subject, and in today's society those experts have less time to devote to these important activities. Nevertheless, the difficulty in finding financial resources for updating many of the reports has been the factor causing the greatest delays. This is true for the nutrient requirement reports on food-producing animals, as well as on laboratory animals, companion animals, and zoological and wildlife animals. Perhaps users of these will more clearly recognize the importance of providing resources for additions of various species and timely updates of existing reports. The importance of these animals to our society dictates that appropriate attention be given this area in order to provide the latest scientific information about the nutrition of these animals to their supervisors and caretakers. For example, it has been about 15 years since the reports on nutrient requirements of cats and dogs have been revised. If one looks at the number of references in the past in these publications (Figure 8-3) there is no reason to believe that research has diminished in companion animals and the importance of companion animals in our society has steadily increased during the past 70 years. Even though the database for companion animals has historically lagged behind that of food animals, there have been hundreds of companion animal nutritional articles published during the past 15 years. It is important that this new information be summarized for the next generation.

RAPIDLY CHANGING RESEARCH AND EDUCATION NEEDS

There are many diverse areas of nutrition that are evolving and that require increased attention to research and education efforts. Several broad areas are briefly described below.

Defining Nutrient Requirements of Animals

To ensure an adequate science base for future generations, nutrient requirements need to be continually refined and extended beyond the current 29 species that CAN addresses. Emerging food and fiber animals, such as ostriches and llamas, also need to be considered. The nutrient needs of laboratory animals are becoming increasingly important in research because their nutrient needs are critical to the interpretation of research results. At some point, a series on nutrient requirements of wildlife and zoo animals will be required to help protect them and prevent them from becoming extinct. As the number of pets increases (current estimates indicate that there are about 56 million dogs and 70 million cats in the United States), so does the need to know their nutrient requirements. The nutritional information for these latter two species will need to emphasize optimal nutrition for health and longevity over that of rapid growth and production that has been so important for food animal production. This will necessarily require more research on nutrition-disease interaction and nutrition and age-related problems.

Protecting Health and Safety

Another emerging research need in nutrition is a better understanding of the pharmaceutical effects of particular nutrients and supplements. Over 7.5 million Americans used St. Johns Wort to treat depression and 7.3 million Americans used Echinacea to treat colds last year. However these substances may have undetermined nutritional and metabolic effects and interactions. There are a host of substances being consumed by humans and their pets that currently have no scientifically proven benefits. Research lags behind use, although a considerable amount of work is now being conducted in some of these areas. In the United States, CAN is a logical authoritative body to pay careful attention to and clarify the role of nutrients, supplements, and "lay" nutritional therapy in animal health.

Animal Welfare

Resolving animal welfare and animal rights issues are also important for future generations. While the quality of animal care has risen during the past 70 years, there is still much to do. It will fall upon the scientists to make major contributions to the health and well being of animals, nutrition being the foundation for the health and welfare of all animals.

Professional Credentials

Because the profession of nutrition has greatly expanded over the years, there is an increasing need to ensure the qualifications of those who provide advice on animal nutrition. Currently, there are no standards or certifying procedures for animal nutritionists. Only 23 states require certification for dietitians, while none require any certification of those practicing animal nutrition. Certification or standard-setting is a much needed element of the professional qualifications of our future animal nutritionists.

CONCLUSIONS

There are several important research and education needs for future generations of scientists, educators, policy makers, and the general public. Nutrient requirements need to be refined and extended to as many species as possible to ensure their health, protection, and longevity. Computer model programs, like the one so elegantly developed in *Nutrient Requirements of Swine* (1998), should continue to be developed for other species. Evaluation of the effects and use of nutrients and nutritional supplements should be based on science. And finally, nature, history and theology should form a moral basis for animal welfare issues and adequate nutrition should continue to be one of its foundations.

REFERENCES

National Research Council. 1986. Nutrient Requirements of Cats, Revised Edition. Committee on Animial Nutrition. Washington, DC: National Academy Press.

National Research Council. 1985. Nutrient Requirements of Dogs, Revised Edition. Committee on Animial Nutrition. Washington, DC: National Academy Press.

National Research Council. 1998. Nutrient Requirements of Swine, Tenth Revised Edition. Committee on Animal Nutrition. Washington, DC: National Academy Press.

9

International Relevance of Feed Composition Information

PHILIP THACKER
University of Saskatchewan, Canada

Some people believe feed resources would be better used if fed directly to humans, instead of animals. However, animal proteins, including meat, milk, and eggs, generally have higher nutritional value than do plant proteins in terms of their amino acid composition and the amounts of minerals and vitamins they supply.

With the exception of vitamin B_{12}, it is possible to obtain a nutritionally adequate diet solely from plant sources, but many people prefer animal protein. Per capita meat consumption in virtually all countries of the world is directly proportional to per capita income.

Vast acreages throughout the world, including arid and semiarid lands, are simply unsuitable for grain production. However, these lands can be successfully used to produce forages. Ruminants can graze forages and convert them to animal proteins. Pigs and poultry do not compete directly with humans for their food supply because feed grains (field corn, grain sorghum, barley and oats) are much different than food grains (rice and wheat).

The livestock industry has also evolved to use byproducts from human industries. For example the crushing industry makes polyunsaturated fatty acid-rich oils for human markets. Byproducts of this industry are canola, soybean, sunflower, and safflower meals, which are used for livestock feeding.

FEED COMPOSITION DATABASE

Literally hundreds of different feeds can be successfully fed to livestock, which raises the need for feed composition databases. Animals require approximately 40 different nutrients to allow them to grow, reproduce, and produce milk, meat, eggs, or wool. Knowing how much of these 40 different nutrients a feed contains is necessary to properly formulate balanced diets. The National Research Council (NRC) publishes and updates the nutrient requirement series of domestic animals. These books are considered the "Bible" of animal feeding and are widely utilized by livestock producers, feed manufacturers, veterinarians, extension agents, and researchers. The back of each of these series contains information on the nutrient composition of feed ingredients particular to the species that it was written for. These tables are produced by various NRC subcommittees and are really the only feed composition tables that are subject to peer review.

Although the NRC publishes feed composition tables, it does not maintain a feed database. In fact, no national feed database exists in North America. Utah State University previously had the only feed database, but it was discontinued in 1990 after being transferred to the National Agricultural Library. Now, the only sources are a few private databases run by companies.

Within the NRC, each subcommittee develops its own estimates of the composition of feeds independently. Each subcommittee compiles as many feed data sets as possible to create an average that is most representative of the feedstuff in question. But this averaged value may not always be correct in all situations.

For example, cereal grains feed compositions are affected by the cultivar, climate, stage of maturity at harvest, the soil composition, fertilizer, and storage practices. Processing techniques that the feed industry uses also alter nutritional composition; so, one feed value will not always be representative of all samples.

Published values are also influenced by the techniques used to analyze the feedstuff. A major problem of building new databases out of old ones is that the older techniques used to analyze the feedstuff were not documented and were often simply not accurate. So, the bottom line is, virtually all of the feed composition values currently used really are not accurate enough. They belong to the horse and buggy age, and it is time to travel to the computer age.

FEED DATABASE CHARACTERISTICS

A North American Feed Database that is flexible enough to address the needs of the livestock industry for the 21st century needs to be developed. A few years ago, the NRC published a book called "Building a North American Feed Information System," which outlined the need for a national feed data base. I urge everyone to read the book and lobby for the re-establishment of a feed database.

A successful feed database must be user friendly, accessible, and accurate. In Canada, the protein content of canola meal has dramatically changed in a 10-year period because plant breeders are continuously improving plant genetics. So, the database must also be flexible enough to change as feedstuffs change.

An estimate of nutrient variability is more important than actual values for various nutrients (see Table 9–1). Therefore, databases should contain an N-value and some sort of standard error associated with that value, so feed companies can build in a margin of safety. History of the sample, growing and processing conditions, and analytical techniques must also be included in the database.

TABLE 9–1. Variations in the protein content of common feed ingredients.

Feed Ingredient	Mean (%)	SD	Range (%)
Barley	11.5	0.91	10.1–13.3
Corn	9.3	0.51	8.2–10.0
Oats	10.4	0.93	9.2–11.6
Wheat	14.7	1.42	11.6–17.9
Soybean meal	46.5	0.49	45.1–47.5
Canola meal	35.4	0.67	34.7–36.4
Blood meal	92.9	1.61	90.4–94.5
Meat meal	50.1	2.18	43.7–59.8

Source: National Research Council series on Nutrient Requirements.

The user should also be able to manipulate and select the parts of the database that they need. For example, the user should be able to determine what the lysine content for the 1998 corn crop was in their area. Feedstuffs change over time, so the database should allow the user to determine which forages have a certain protein content.

BENEFITS OF A NORTH AMERICAN FEED DATABASE

A computer based feeding system is estimated to cost between $1.5 and $2.0 million per year to operate and maintain (National Research Council, 1995). Producers are not currently in a good position to fund that kind of a database. The reality is that all society would benefit from a feed database; so, maybe the entire society should pay for it.

If feed nutrient composition values are known, then animal production and the food supply will be cheaper. For example, daily gain, feed conversion, and preweaning mortality can all be improved by animal nutrition. One feed integrator estimates that for an 800-sow operation, if feed conversion is improved 0.1 of a unit, then $27,000 could be saved. If daily gain is improved 25 g/day, then $31,000 could be saved. And, if preweaning mortality rates are decreased from 11 to 10 percent, then $70,000 could be saved.

Producers use a margin of safety because they are not confident about the nutrient content of feedstuffs they use in diet formulation. Accurate information about nutrients used in feed will enable producers to accurately formulate diets to meet the animals requirements, which will reduce emissions from livestock operations.

A feed database could also catalog anti-nutritional factors to avoid toxicity problems and ensure feed and food safety. The database could also improve national trade by providing accurate nutritional composition.

CONCLUSIONS

A dynamic, computerized feed database is needed because tables used in the past are static, are no longer accurate, and are not flexible enough for the feed industry of the 21st century. Computerized databases that change as feed changes and allow the user to manipulate variables will make animal production more efficient and provide consumers with a safer and cost-effective product.

REFERENCES

National Research Council. 1995. Building a North American Feed Information System. Washington, DC: National Academy Press.

10

The International Aquaculture Market and Global Needs

DANIEL VILLAMAR
Cargill

Author's present address: AcuaBiotec LLC, 10400 Windfall Court, Damascus Maryland 20872 USA

Aquaculture production is meeting a rising demand for a variety of fishery products, including fish, shrimp, shellfish, and aquatic plants. The capture fishery for seafood used for human consumption has remained stable at about 60 million metric tons per year from 1984 to 1996. However, aquaculture production has increased from about 13 to over 35 percent of the capture fishery yield over the same period, exceeding 20 million metric tons per year in 1996 (Figure 10-1) (Tacon, 1998a,b; New, 1997). During this time, major commercial fishing grounds have been classified as "fully exploited" or "over exploited." The Food and Agriculture Organization (FAO) estimates world annual consumption of seafood in year 2010 to be 110 to 120 million metric tons with an estimated supply of 74 to 114 million metric tons. A likely scenario to expect by the year 2010 will be a deficit of 36 million to 46 million metric tons of seafood for human consumption (Food and Agriculture Organization, 1996). With capture fisheries seemingly limited to 60 million metric tons per year, aquaculture production must double over the next 15 years simply to keep

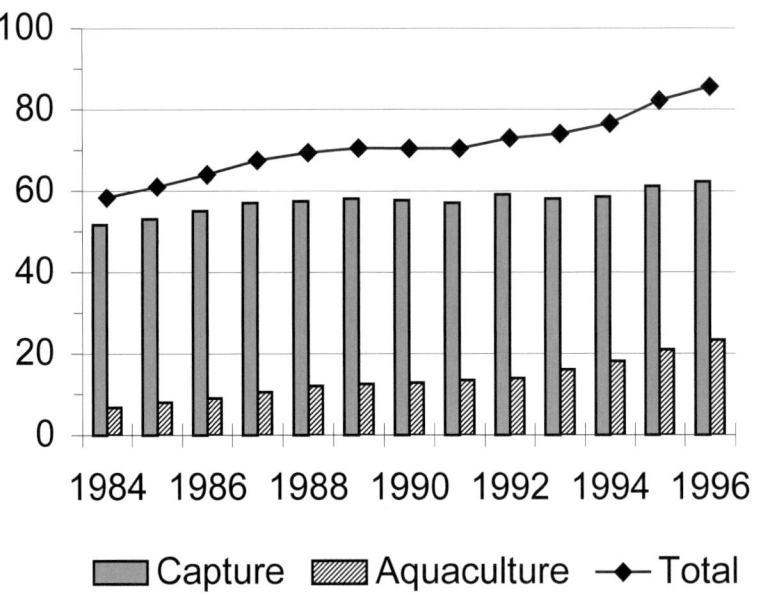

FIGURE 10–1. Millions of metric ton (t) of capture and aquaculture fisheries production for human consumption.

up with expected population growth at today's per capita seafood consumption. Over 120 species of fish are farmed throughout the world, and the diversity of species is nearly matched by the diversity of farming practices used to produce them.

OVERVIEW OF WORLD FISH PRODUCTION

The majority (nearly 90 percent) of fish production from aquafarms occurs in Asia, especially China. China is the only country where aquaculture supplies food for more people than does the country's capture fishery industry. In 1978, China produced less than 1 million metric tons of fish through aquaculture and 2.5 million metric tons through its capture fishery industry. But in 1997, China produced 20 million metric tons of fish from aquaculture and 15.7 from capture fishery (Fish Farming International, 1998).

China raises a great variety of aquatic animals for human consumption ranging from turtles to crabs to eels. However, production of cyprinids, carp, and related freshwater fish species, which are an important and inexpensive protein source consumed internally, comprise the greatest aquaculture crop in the world (Tacon, 1998a,b; New, 1997). At approximately 10 million metric tons and $10 billion dollars in value, cyprinid production, accounts for about 76 percent of the world's total aquaculture production tonnage and 46 percent of the world's aquaculture crop value. At about 0.5 million metric tons and approximately $3.5 billion in value, tiger prawn—the most popular shrimp species grown in Asia—accounts for about 4 percent of the world's tonnage and over 16 percent of the world crop value.

Salmon farming has become a major industry in Norway, Scotland, Chile, and Canada, and today, over 70 percent of the salmon consumed in the world comes from salmon farming, rather than from capture fisheries. Salmon production tonnage is similar to that of tiger prawn and tilapia and ranks third in terms of value at approximately $1.8 billion. The top species grown in aquaculture and their market value are represented in Figure 10-2 (Tacon, 1998a).

Shrimp farming supplies about one quarter to one third of the world's consumption and is a major source of foreign exchange for many developing countries, with Thailand and Ecuador as number one and two shrimp-producing countries over the last decade, respectively. About 80 percent of the world's farmed shrimp is produced in Asia, and the remainder is produced primarily in Latin America. The United States and Japan are the world's largest importers of shrimp. Farmed shrimp production has a world trade value estimated at about $6 billion from about 814,000 metric tons of live weight produced in 1999 (Rosenberry, 1999). This represents an increase of about 10.4 percent over 1998 production. The increase came from Asia, as the major producing countries in Latin America suffered severe losses to diseases in 1999, and these losses continued in 2000.

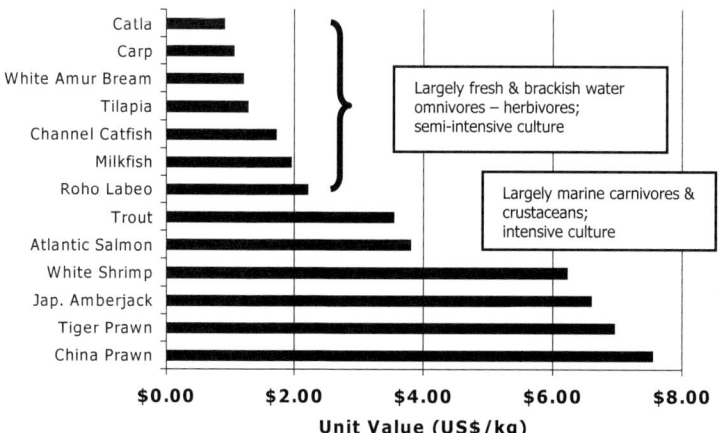

FIGURE 10-2. Top species by unit value (Tacon, 1998b)

The most destructive diseases have affected the world's shrimp farming industry for over a decade, with the first massive losses experienced in Taiwan in 1987-88 and then continuing through Asia in the 1990s and crippling the West in the last two years. These severe epizootics include several different viruses, which appear to cause the most damage when they occur in combination with pathogenic *Vibrio* bacteria. Estimates of some of the losses and economic damage caused by major epizootics are presented in Table 10-1 and losses realized by Thailand and Ecuador are presented in Figure 10-3.

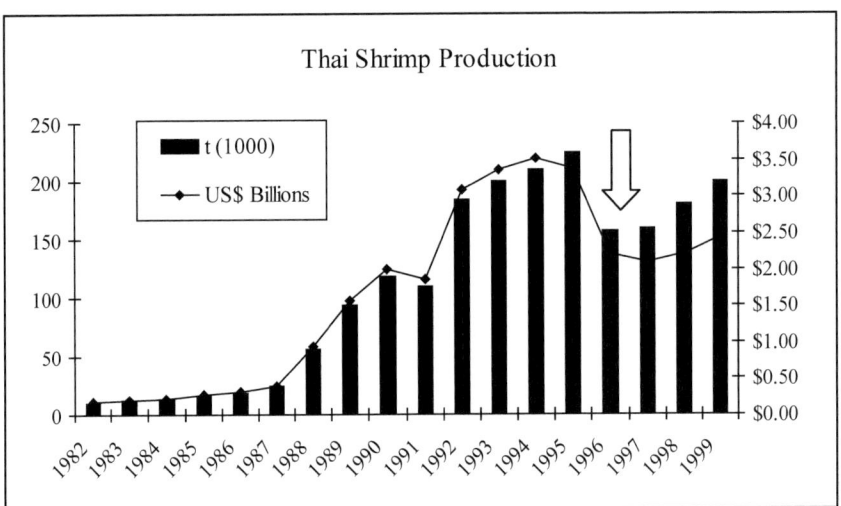

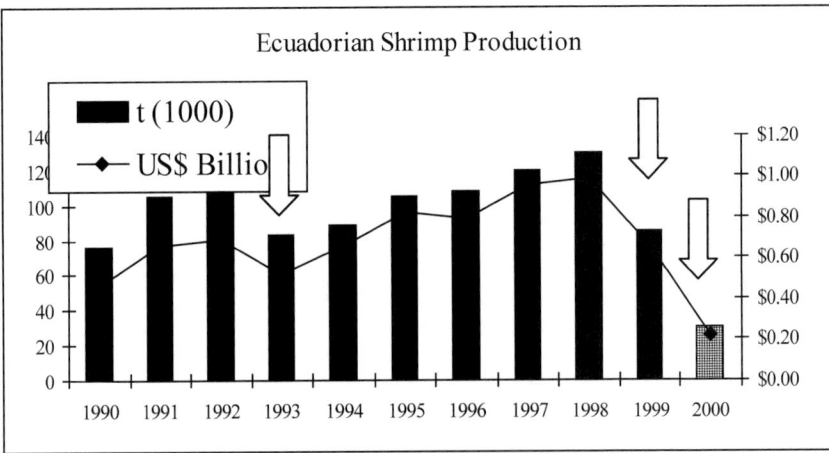

FIGURE 10-3. Severe effects of disease on shrimp production in Thailand and Ecuador. Arrows indicate decline in production due to disease. Year 2000 values are projections.

TABLE 10-1. Some estimated losses to disease in world shrimp farming.

Country	Year	Disease	Loss (Tons)	Loss (%)	Loss (Billions US$)
Taiwan	1987-1988	Vibrio, virus[a]	60,000	75%	$0.60
China	1992-1994	WSSV[b], Vibrio	150,000	75%	$1.20
Ecuador	1992-1994	Vibrio, TSV[c]	24,406	22%	$0.15
Ecuador	1999	WSV[d], Vibrio	50,000	38%	$0.27
Thailand	1995-1996	WSV, YHV[e], Vibrio	67,500	30%	$0.95
Total			346,906		$3.17

Suspected or confirmed diseases agents include: [a]white spot virus (suspect); [b]white spot syndrome virus; [c]taura syndrome virus; [d]white spot virus; [e]yellow head virus.

Application of chemicals and antibiotics to aquatic feeds and water are the most commonly used methods to fight diseases, especially in shrimp farming. In fish such as salmonids, which have a specific immune system, vaccines can be used to prevent disease. However, shrimp do not have a specific immune system, and the widespread use of chemicals and antibiotics, which are ineffective in fighting viral diseases, are harmful in the long term. Strong microbiologic evidence indicates that with the use of antibiotics and chemicals, bacterial pathogens that cause disease in humans, such as *Vibrio cholerae*, have evolved (horizontal gene transfer) to more harmful and more resistant types (De La Cruz and Davis, 2000; Rowe-Magnus and Mazel, 1999). By implication we can surmise that the same has occurred in aquaculture; in regions where antibiotic use has been heavy, bacterial pathogens now cause greater rates of shrimp mortality than in previous years (Moriarty, 2000). Continued use of antibiotics and chemicals in open systems can cause serious damage to the aquaculture industry in the long-term.

AQUACULTURE PRODUCTION PRACTICES: A PYRAMID

Aquaculture production practices have been represented as a pyramid, serving as a visual aid to categorize the broad ranges and levels of technology inputs practiced around the world (Figure 10-4; Tacon, 1988). At the base of the pyramid are relatively inexpensive aquatic animals produced under extensive conditions, (i.e., in large pond areas stocked at low densities with low inputs in terms of husbandry and nutrition technology) comparable to pasture production of cattle. Inexpensive organic material such as manure is used as a fertilizer to stimulate natural production as forage for the aquatic livestock, which are primarily herbivorous and omnivorous fish such as carp or tilapia or fresh water

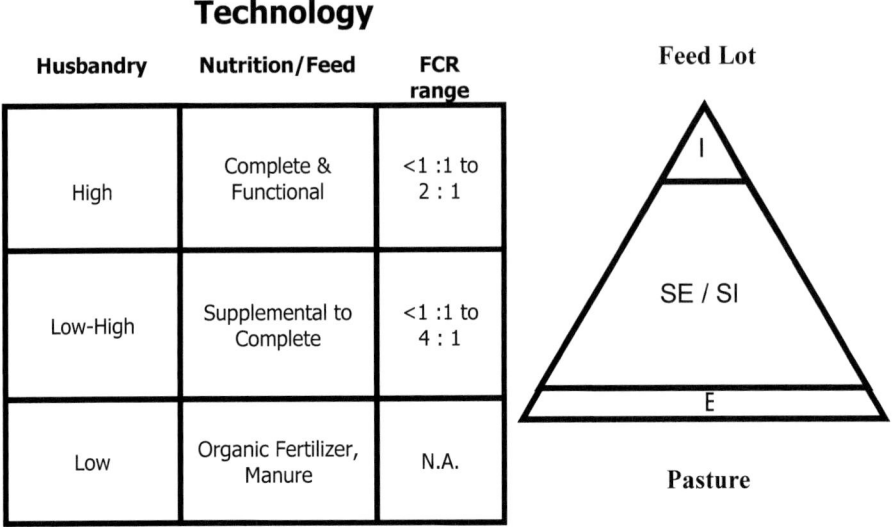

FIGURE 10-4. Intensive, semi-extensive/semi-intensive and extensive aquaculture systems. (Tacon, 1988).

crustaceans. In these extensive systems, seedstock can be purchased at low cost from fishermen or from local hatcheries or simply brought in by tidal action. Overall inputs, in terms of financial investment, levels of technology and skills required are very low, as are the associated risks.

The large volume in the center of the pyramid represents semi-extensive to semi-intensive production of aquatic livestock, which includes a broad range of species (omnivores, herbivores, carnivores) raised under a broad range of conditions. Inputs in terms of husbandry technology range from low to high, as does the relative sophistication of nutritional programs and required feeds. Farmers tend to use supplemental feeds and depend to varying degrees on the nutrition provided by natural food organisms to help sustain their aquatic livestock. Feed conversion ratios (FCR) can range from less than 1:1, where farmers take advantage of natural food contributions, to 1:4 or greater, in situations where supplemental feeds are largely inadequate to meet animal dietary requirements and/or where natural food production is limiting.

The peak of the pyramid represents intensive aquaculture, similar to intensive animal production, such as cattle feedlots or poultry production houses. Feeds used in intensive systems are usually complete in terms of meeting the animal's nutritional requirements throughout the production cycle, and functional with respect to the required physical-chemical characteristics, such as water stability, buoyancy and organoleptic properties. Complete and functional feeds are used to raise mostly marine carnivores, crustaceans and other high-value fish with a range in FCR of less than 1:1 for salmon to about 2:1 for other species. The very low FCR seen in salmon reflect a high level of nutrition and feed manufacturing technology in synch with sophisticated husbandry practices.

In the following descriptions of aquaculture operations, the convention of the pyramid is used to highlight relative position in the range of practices in aquatic farming.

Extensive Production Systems

Aquaculture in China is incorporated into everyday life with fish production ponds located near population centers and easily accessible to the marketplace. Traditional Chinese fish production systems are often integrated with farm animal production (e.g., manure from swine pens is used to fertilize fish ponds), thereby stimulating productivity of the aquatic food chain (Figure 10-5). In another type of extensive fish production system, aquatic plants are grown in one pond and used to feed grass carp in another (Figure 10-6). Likewise in Indonesia, large fleshy leaves are fed to herbivorous fish. These fish are aggressive eaters and quickly reduce leaves to veins (Figure 10-7).

FIGURE 10-5. Extensive production system: Use of animal manure for fertilization.

FIGURE 10-6. Extensive production system: Harvesting food for fish.

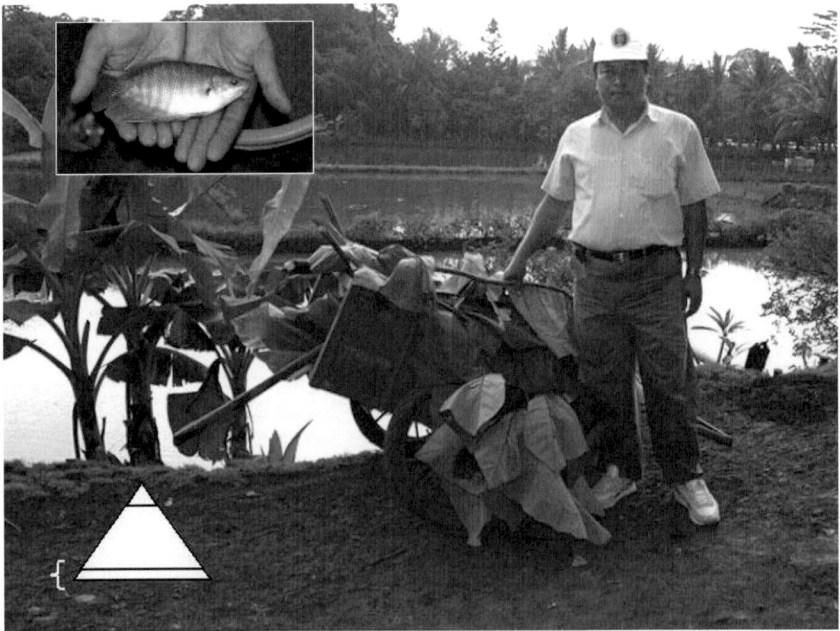

FIGURE 10-7. Extensive production system: food fed to fish.

Semi-Extensive Production Systems

Semi-extensive carp production is practiced in Poland and other parts of Europe with large ponds, stocked at low densities and use of whole grains as feedstock. Interestingly, pond production of the common carp has been practiced for so long in these areas that carp have been selected to have very few scales, reportedly because housewives successfully argued that they did not like to scale the fish. Feed inputs limited to whole grains result in a feed conversion ratio of over 3.5:1 in semi-extensive carp farming (Figure 10-8).

As an example of semi-extensive aquaculture practice used to raise shrimp, farmers in Vietnam use tidal action to bring in water to their ponds, not requiring the use of pumping stations, and this same natural influx of water can supply shrimp seedstock for the ponds. At the low stocking densities, less than five shrimp per square meter of surface area, natural productivity contributes strongly to the nutrition of the shrimp and there is little need for complete feeds. Shrimp are fed a locally made mixture of boiled trash fish and rice bran (Figure 10-9). Fresh water prawns are farmed in a similar manner in Thailand and fed supplemental, homemade feeds consisting of sun dried, cold-extruded, noodles made of rice bran and trash fish. In most of Latin America, where approximately 20 percent of the world's shrimp are farmed, production systems are semi-intensive; farmers use diesel-powered or electric pump stations to supply water to large ponds, purchase seedstock from hatcheries or from local fishermen, and purchase supplemental and/or complete pelleted feeds that are commercially manufactured.

FIGURE 10-8. Semi-extensive production system for carp in ponds.

FIGURE 10-9. Semi-extensive production system: locally made food mixture.

Intensive Production Systems

At the top of the pyramid are the intensive fish/shrimp farming systems, where nearly all production parameters and technology inputs are well controlled, including type and quality of seed stock (fry or postlarvae) and feed. In the case of intensive shrimp farming, postlarvae are purchased from commercial hatcheries where selective breeding may have started and where sophisticated laboratory techniques are used to screen for viral and other diseases. Intensive shrimp feeds are nutrient-dense, complete, and functional. For example, in Thailand, shrimp are farmed in small, uniform ponds less than one-hectare in size with heavy mechanical aeration used to maintain adequate concentrations of dissolved oxygen and to circulate water helping to keep organic waste in suspension and aiding its aerobic decomposition (Figure 10-10). Stocking densities are high, with yields of 6 metric tons per hectare per crop or greater. Intensive feeds that cost over $1.00 per kilogram can provide an FCR of about 1.4 to 1.7 :1. These feeds are sold as small, highly compact, water stable pellets made of very finely ground ingredients, including marine proteins, sources of cholesterol, lecithin, attractants, and pigments. Feed trays are used to monitor feed consumption and shrimp health throughout the day. Phase-feeding is practiced to improve rate of weight gain, and water exchange is minimal to help control diseases. Shrimp are harvested quickly by draining the small ponds and are immediately available for processing maintaining excellent product quality (Figure 10-11).

Intensive fish farming can be seen in Malaysia, where concrete ponds filled with tilapia are supplied with oxygen by aerators (Figure 10-12). Feeds used are nutritionally complete, floating, and extruded, contain about 30 percent crude protein and 5 percent fat, and can achieve an FCR of about 1.5:1. In many countries, floating cages are used to raise high-value marine fish such as sea bass and sea bream. Where manufactured feeds are not yet available, ground trash fish is used as feed. Other species, such as hybrid catfish and snakehead, are also intensively farmed in Asia. In Thailand, hybrid catfish are fed complete, floating, extruded feeds containing about 35 percent crude protein and, 6 percent fat with an FCR of about 1.2:1. Most of the hybrid catfish production is for domestic consumption rather than export, as is the case for shrimp.

Other intensively reared species of fish include salmon, trout, European sea bass, sea bream, yellowtail, catfish, and eel. All of these species receive complete, pelleted feeds. With the exception of channel catfish feeds, the feeds for these species contain substantial concentrations of fish meal. Feeds for aquatic species now account for over 25 percent of global fish meal production, up from less that 10 percent one decade ago. Given than fish meal production has been stable over the past decade, averaging 6.5 million metric tons per year, expanded production of these species will depend upon the development of alternative sources of protein from sources that can be increased in the future or diverted from other uses. This suggests that development of suitable protein sources from grains and oilseeds will play an important part in expanded production of intensively farmed fish species.

Four years were required for salmon to reach harvest size a decade ago; today, it takes less than two years for salmon to reach the same harvest size. This gain is the result of improvements in feed formulation, manufacture, and feeding practices, plus the development of vaccines that prevent infectious diseases. Similar advances are being made in the other species of intensively farmed fish.

FIGURE 10-10. Intensive production system: phase feeding.

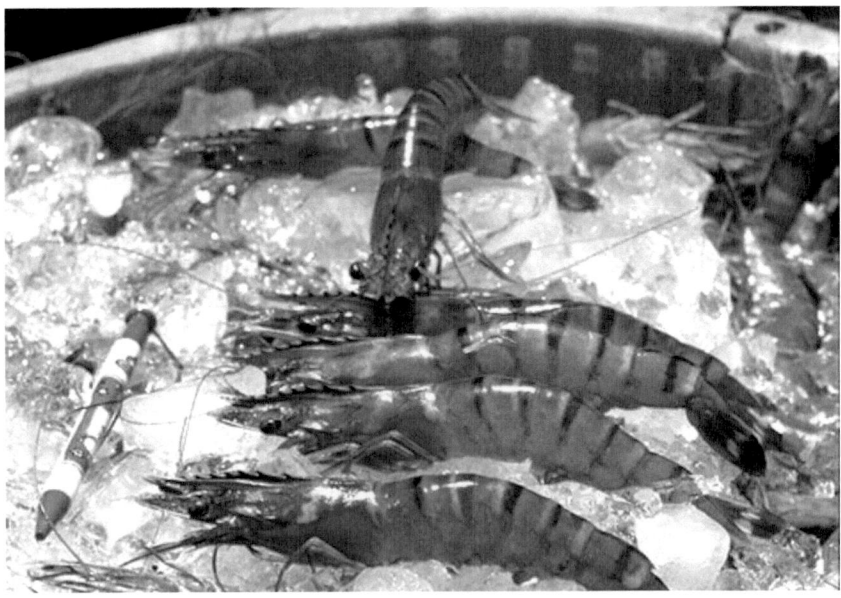

FIGURE 10-11. Intensive production system: quality product harvesting.

FIGURE 10-12. Intensive production system: aeration.

INDUSTRY NEEDS

The aquaculture industry requires more rapid development in three principle areas of technology required to support sustainable growth: nutrition, genetics and environment. In terms of nutrition and feeds, better understood nutritional requirements of fish and clearly understood production requirements of farmers are needed because of the great diversity of fish species used by the aquaculture industry. Appropriate technologies in feed formulation, manufacturing, and feeding are needed to help the industry parallel modern farm animal production systems and facilitate the transition from extensive to semi-intensive and intensive systems where natural resources can be more efficiently used.

In hatcheries, the dependence on live and fresh food organisms, especially microalgae and brine shrimp (*Artemia*) used to raise larval shrimp and fish must be replaced by commercial feeds to help control disease vectors and maintain consistent production. The world's supplies of *Artemia* cysts have been in great decline, and high prices have made existing supplies unavailable to many farmers. Conventional larval feeds, including dry microcapsules, flakes, and powders, decompose rapidly in water, increasing rates of microbial loading, disease, and mortality of larval shrimp and fish. Liquid feeds, which consist of fully hydrated microcapsules stabilized in a probiotic liquid medium, are a new category of larval feeds made by manufacturing methods that prevent excessive loss of nutrient bioavailability. Liquid feeds are superior to conventional dry larval feeds, reduce pollution, and can replace a greater proportion of live food organism used in hatchery feeding programs. Complete replacement of live foods remains a high priority for the industry. For grow-out fish and shrimp, more efficient feeds are needed, especially those that lower the amounts of enriching nutrients (e.g., nitrogen and phosphorus) that are excreted into the aquatic environment and contribute to eutrophication of waterways.

Strong emphasis must be placed on stock domestication and selective breeding programs to provide more predictable and consistent animal performance, especially in the case of shrimp. Most shrimp hatcheries continue to depend on wild-caught broodstock, which can carry diseases and can have low or variable fecundity and unpredictable animal performance. Efforts to develop disease tolerant or resistant strains of shrimp are underway in many countries, but ever-increasing varieties of new viral and bacterial pathogens make it difficult to keep up. Genetic improvements of most aquatic species are needed to help move the industry forward.

Advances in nutrition and genetic technologies must be accompanied by similar advances in bio- and other technologies to help control the aquatic microbial environment, especially in shrimp farming. That is, if the environment is not under control with respect to microbial ecology, then technologic advances in nutrition and genetics will not yield full benefits. The proliferation of more virulent bacterial and viral pathogens requires new and better approaches in the control of disease. This means a move to closed systems, which has already started in several countries, and a focus on disease prevention, not eradication, by managing microbial communities to maintain pathogens at levels that permit operation of profitable businesses (Moriarty, 2000).

CONCLUSIONS

Aquaculture presents global challenges, including great diversity in species, habitats, culture systems, and industry technology. World markets are diverse. In addition, the nutrient requirements of most fish species and the bioavailability of nutrients in many locally produced ingredients are not known.

The aquaculture industry needs to eliminate its dependence on capture fishery, both as a source of seedstock and as a source of feed ingredients, primarily fish meal. Feed programs need to be based on nutrient content rather than on ingredients. Closed aquaculture systems utilizing recirculated water must be developed to the point where production costs are competitive with systems depending upon flowing water. Domesticated strains of fish selected for economically important traits must also be developed. For some species of farmed fish, the sophistication of farming is approaching that of the poultry industry, but, for most others, fish farming is decades behind in terms of controlling inputs and outputs such that farming is sustainable, environmentally benign, and economic. Technologic advances in genetics, nutrition and environment, must be made in parallel to achieve long-term industry sustainability.

REFERENCES

De La Cruz, F., and J. Davis. 2000. Horizontal gene transfer and the origin of species: lessons from bacteria. Tr. Microbiol., 8(3): 128-133.

Food and Agriculture Organization. 1996. The State of the World Fisheries and Aquaculture.

Fish Farming International. 1998. China output exceeds forecasts. 25 (11), pp. 1, 40-41.

Moriarty, D.J.W. 2000. Disease control in shrimp culture with probiotic bacteria. In Bell, C.R., Brylinsky, M. and P. Johnson-Green (eds.), Microbial Interactions in Aquaculture. Proceedings of the 8th International Symposium on Microbial Ecology, Atlantic Canada Society for Microbial Ecology, Halifax, Canada, 2000. pp.237-243.

New, M.B. 1997. Aquaculture and the capture fisheries. World Aquaculture, June 1997. pp.11-30.

Rosenberry, B. 1999. World Shrimp Farming 1999. Vol. 12. Shrimp News International, San Diego California.

Rowe-Magnus, D.A., and D. Mazel. 1999. Resistance gene capture. Curr. Opin. Microbiol., 2:483-488.

Tacon A.G.J. 1998a. Global trends in aquaculture and aquafeed production 1984 –1995. In Fraser S (Ed.), International Aquafeed Directory and Buyers' Guide 1997/98. Turret RAI PLC., Middlesex U.K. pp. 5-37.

Tacon A.G.J. 1998b. FAO Aquaculture production update. In Fraser S (Ed.), International Aquafeed, Issue 2. Turret RAI PLC., Middlesex U.K., pp. 13-16.

Tacon, A.G.J., 1988. The nutrition and feeding of farmed fish and shrimp-a training manual. 3 Feeding Methods. FAO Field Document, Project GPC/RLA/075/ITA. Field Document 7/E, Brazilia, Brazil. 208 p.

Summary

Meeting the Challenges of the New Century

DALE E. BAUMAN
Cornell University

Discussions presented in this volume highlight some of the significant advances experienced in animal nutrition over the past 7 decades. The discussions also outline important roles of animal nutrition for the future. Our nation faces many challenges in the new millennium. From environmental pollution to products of biotechnology, we rely on a science-based analysis of the issues. Solutions to many of the wide range of animal-related challenges require a science-based application of animal nutrition.

When the Committee on Animal Nutrition (CAN) began its work over 70 years ago, half of the U.S. population lived on farms. Our knowledge of nutrition was extremely limited, and the challenges were very different. Only a few vitamins and minerals were identified, and essential amino acids and essential fatty acids were not understood. Progress during this period in developing an understanding of nutrition came about predominately through trial and error.

Today, over 50 percent of the U.S. population live in cities over 1 million in size. This presents exciting challenges, because fewer people understand agriculture, food production, or animal nutrition. However, knowledge of basic biology is advancing at an exponential rate. Thus, research in animal nutrition can be largely based on knowledge of biological principles and fundamental concepts rather than a trial and error approach.

Food products derived from animals are important components of human food production throughout the world. The CAN reports have played an important role in providing scientific-based information on the nutrient requirements of animals and the nutrient composition of feedstuffs. Reports have been translated into five languages, and their use represents a key element in providing for the needs and well-being of animals. Their present value and future potential can be broadly divided into five topics - sustainable agriculture, food safety and public health, animal welfare, environmental quality, and international trade and development.

SUSTAINABLE AGRICULTURE

Agriculture is and will always be a vital industry because of our need for a safe, high quality, affordable food supply. Animals and knowledge of animal nutrition are key components to an integrated farming system. At the individual farm level, animal nutrient needs are of critical importance in the development of whole farm nutrient management programs. At the industry/farm level, animal use of nutrients represents a key component in utilizing food industry byproducts and the formulation of a "system" nutrient management program.

The CAN reports provide the latest nutritional information so that various feedstuffs and byproducts can be considered on an individual farm basis and utilized in formulating balanced diets for livestock. For example, even now, 25 percent of poultry and swine diets and 50 percent of Florida dairy cattle diets are byproducts from human food and fiber industries. In fact, a few years ago Anheuser Busch proudly advertised that their distiller's grains were used in the diets of ruminants, which represented an effective part of their waste recycling program. The use of byproducts as animal feeds is expected to increase in the future and will continue to be an important component in the food industry's nutrient management programs.

FOOD SAFETY AND PUBLIC HEALTH

Feeding animals appropriately improves their ability to produce high quality food products. The composition and nutritional quality of milk, eggs, and meat can all be affected by the diet the animal consumes. For example, if an animal is fed a diet that is inadequate in certain vitamins, then the vitamin content of food products from those animals will also be below normal.

When animals are adequately nourished, their resistance to disease also improves, which reduces the potential for animal pathogens that might adversely affect public health. Improving animal health through nutrition also decreases the need for drug treatments, which reduces the potential of drug residues. Thus, feeding animals adequate amounts of a well-balanced diet represents an important consideration that impacts our food safety assurance system.

Niche markets and organic foods are developing areas in food production. In some cases, the specialized requirements for niche markets raise special challenges in meeting the animal's needs and providing for its well-being. Practices used to create these foods need to carefully consider the animal's nutritional requirements and the nutrient value of feedstuffs to provide adequately for the health and welfare of the animals and ensure quality of the food products is maintained.

Future changes in animal care, modifications in housing design, and improvements in management systems will all impact nutrient requirements. Food production systems are gradually shifting to larger, more concentrated operations, and this offers new challenges. For example, this presents the potential for a biohazard to involve a much greater number of animals. A sound nutrition program is essential to decrease the potential for rapid spread of disease among animals on a single farm or many farms. Thus, applying current knowledge of nutrition to optimize animal health and disease resistance is essential.

Basic nutrient requirements of animals used in food production will also need to be continually evaluated as scientists create "designer foods." Identifying microcomponents of foods associated with beneficial health effects is a growing emphasis often referred to as "functional foods." An understanding of the biology in this area will allow researchers to make modifications to increase the concentration of these microcomponents in food products derived from animals. One example is formulating diets that enhance the concentration of omega-3 fatty acids in animal products. Another example is to feed and manage animals so that concentrations of conjugated linoleic acid (CLA) in milk and meat are increased. CLAs are potent naturally occurring anticarcinogen and the National Research Council has pointed out that they are the only fatty acids known to unequivocally inhibit cancer in animal models.

ANIMAL WELFARE

Predicting animal nutrient needs more precisely improves animal welfare. In particular, it allows better diets to be provided to animals, which is beneficial to well-being and disease resistance. The CAN reports provide users with a better understanding of nutrient requirements and will continue to advance the care of food animals, exotic animals, laboratory animals, and companion animals.

Improving diet quality is critical for the conservation of threatened or endangered species. Proper nutrition is essential for growth and well-being, as well as normal reproduction and perpetuation of the species. In fact, Charles Darwin, in his classic research on the origin of the species, was among the first to recognize that reproduction was severely compromised when the food supply was inadequate. In addition to the supply of food, a correct nutrient balance is essential, and identifying these needs in exotic animal species is a rapidly evolving area.

Animals used in any type of research, whether at universities, medical laboratories, the USDA, the NIH, or in outer space, need to have adequate nutrition to ensure tests of biological hypotheses are valid.

Animals providing public service such as "seeing eye" or military dogs need to have proper care. Americans are increasingly conscious of the importance of nutrition in maintaining their own health and the health of their companion animals. CAN reports are a critical resource for veterinarians so that proper nutrition recommendations can be made to their clients.

ENVIRONMENTAL QUALITY

The goal for producers and farmers is to carefully manage animals so that a high productive efficiency is achieved with minimal environmental impact. Feeding animals adequate amounts of a well-balanced diet will allow the animal to maintain productive efficiency while reducing the excretion of excess, unutilized nutrients that may have adverse environmental impacts. For example, reducing animal waste content of nutrients such as nitrogen, phosphorus, copper, and selenium is beneficial in protecting soil and water quality and enhancing the sustainability of animal agriculture.

Animal nutrition must be a central component of the Environmental Protection Agency's guidelines for animal production systems. And CAN has a tremendous amount to offer. Formation of guidelines and standards requires scientific evaluation and recommendations, together with a balancing of the needs of industry and the environment. It is important for the Environmental Protection Agency and National Research Council to work together and share expertise in animal agriculture to ensure that environmental pollution is minimized.

INTERNATIONAL TRADE AND DEVELOPMENT

The potential for growth in animal exports is well recognized because of recent trade liberalization created by the North American Free Trade Agreement (NAFTA) and the General Agreement on Tariffs and Trade (GATT). The world market for animal products is expanding, and the competitiveness of the United States will depend on meeting nutrient needs to optimize animal productivity.

Reports by CAN on nutrient requirements are used throughout the world. Thus, the series of CAN reports have played a central role in global transfer of nutrition-related technology. US researchers visiting developing countries often take along CAN reports, because they are eagerly sought by scientists and applied in their countries. The ability to predict an animal's nutrient requirements under varying environmental conditions and resource availability enables producers around the world to better manage and care for their livestock.

FINAL THOUGHTS

While CAN reports have been of tremendous value, the shifting paradigms of technology offer exciting challenges and opportunities for the future. Dynamic models of animal digestion and metabolism allow greater precision in developing animal requirements and make the reports of even greater value. The swine publication is the first National Research Council publication of this type, representing a dynamic metabolism model based on amino acid ratios. As a result, amino acids are used more efficiently with less amino acids oxidized and less nitrogen excreted in the animal waste. The new dairy report has a dynamic model of rumen digestion. The challenge with ruminants is to optimize ruminal degradation rates for dietary protein and carbohydrate fractions, to allow maximal utilization by the rumen microbes. Other advancements will undoubtedly lead to improved measures of animal well-being and CAN reports will play an important role in their application. For example, research has shown how nutrient status impacts the endocrine system in the modulation of the somatotropin/insulin-like growth factor axis. These interactions involve a key role for specific nutrients in the regulation of gene expression for processes associated with maintenance of animal well-being.

Recent developments in biotechnology and their application to animal agriculture require that nutritional implications be continuously assessed. For example, lysine is often the limiting amino acid in feedstuffs used for animal diets. Recombinant DNA technology can produce enzymes that could be used as dietary supplements to increase the bioavailability of lysine. Genes for lysine biosynthesis can also be inserted into microbes that become microbial fermenters to produce crystalline lysine for use as a dietary supplement. Other groups of scientists are working to enhance plant lysine content, and genetic engineering also could be used to produce farm animals with the ability to synthesize much of their own lysine requirement. All of these examples are actively being investigated. Each impacts the nutritional requirements of animals, but in slightly different ways. Biotechnology also offers the opportunity to develop other dietary additives. For example, if phytase can be added to diets to improve phosphorus availability, then phosphorus losses in animal waste could be reduced. Additional dietary additives that can be produced by recombinant DNA technology have the potential to enhance rumen microbial fermentation processes thereby increasing the animal's productive output per unit of resource input.

CONCLUSIONS

The Committee on Animal Nutrition of the National Research Council's Board on Agriculture and Natural Resources has overseen the preparation of reports relating to over two dozen animal species. These nutrient requirement reports will continue to be critical in scientific investigations, serving as the reference basis in the conduct of public and private research with animals. The reports are also valuable for educating the next generation of students and industry personnel about the science of nutrition and the principles of a sound animal nutrition program. Further, CAN reports will continue to be a central component in outreach programs to improve the care of farm animals and companion animals. Extension agents indicate that providing an adequate and balanced diet continues to be a major element of feeding programs where compromises occur in the management and care of animals.

CAN has 70 years of impressive accomplishments and successes. This is truly a record that CAN participants, the Board on Agriculture and Natural Resources, and the National Research Council can be proud. The science of nutrition is evolving and expanding in its relevance and importance. Sustainable agriculture, food safety and quality, animal care and well-being, environmental quality, and international trade and development are all areas where the reports from the National Research Council's CAN have played a vital role. Clearly, CAN reports will continue to be a key element in the efficiency, health, and well-being of animals and humans throughout the world.

Authors

ABOUT THE AUTHORS

DALE E. BAUMAN is Liberty Hyde Bailey Professor in the Department of Animal Science at Cornell University. He received his undergraduate and Master's degrees from Michigan State University and his Ph.D. degree in nutrition-biochemistry from the University of Illinois. Prior to his appointment at Cornell University, he was an associate professor at the Department of Dairy Science at University of Illinois. Bauman's research interests include biochemical and hormonal regulation of nutrient utilization for growth, pregnancy and lactation, nutrition and metabolism of ruminants, mammary gland biology, mechanisms of somatotropin, biology of food-producing animals, and animal agriculture. Bauman and his colleagues crystallized the concept of homeorhesis, the process of long-term regulation of nutrient use during a particular physiologic state such as lactation. His concepts of metabolic regulation are widely accepted and applied to many aspects of developmental biology. In 1988, Bauman was elected to the National Academy of Sciences. He served as chairman of the Board on Agriculture from 1994 to 1997 and was a member of the Board from 1990 to 1994. His service to the Academy has included membership on numerous committees, among them most recently, the oversight commission for *Ensuring Safe Food from Production to Consumption* and the authoring committee for the report, *Metabolic Modifiers: Effects on Nutrient Requirements of Food Producing Animals*.

ROBERT J. COLLIER received his B.S. degree in Zoology from Eastern Illinois University in 1969. After service in the Army Medical Corps, he obtained his Master's degree in zoology from Eastern Illinois University in 1973 and his Ph.D. in physiology from the University of Illinois in 1976. His dissertation research was on the endocrine regulation of lactogenesis in the dairy cow. In January 1976, Collier accepted an NIH postdoctorate in the Dairy Science Department of Michigan State University in the laboratory of Allen Tucker. His research was on the regulation of cortisol uptake in mammary tissue of cattle. In September 1976, Collier joined the Dairy Science Department at the

University of Florida as an Assistant Professor, where he developed a teaching and research program on the environmental physiology of the dairy cow in the subtropics. He also continued his research on the endocrine regulation of lactation in cattle as well as swine. He was promoted to Associate Professor in 1981. In July 1985, Collier joined the Monsanto Company as a Science Fellow and initiated a discovery program in lactation and growth regulation. He was promoted to Dairy Research Director and Fellow in 1987 and until 1999 was Dairy Research Director and Senior Fellow. In that capacity, Collier was responsible for all preclinical and clinical research in North America required for the commercialization of bovine somatotropin as well as research on novel factors regulating growth, development, and lactation of domestic animals. Since 1987, Collier has been an Adjunct Professor of the Dairy Science Department at the University of Missouri. Since 1999, Collier has held a position with the University of Arizona. In 1990, Collier was appointed an Honorary Fellow of the Hannah Research Institute, Ayr, Scotland. In 1991, he received the ADSA Upjohn Physiology Award, and in 1992 he was selected as Alpha Omega Alpha visiting professor at the University of Indiana and Donald Barron Visiting Professor at the University of Florida. He has served on the Biotechnology Advisory Board for the University of Iowa and both the Nutritional Sciences Advisory Committee and the Animal Sciences Advisory Board for the University of Illinois. Presently, Collier chairs the College of Science Advisory Board for Eastern Illinois University. He is author or co-author of 136 journal articles, chapters, and reviews, 99 abstracts, 28 popular articles, and 6 U.S. Patents.

DANNY G. FOX is Professor of animal science at Cornell University. He received his B.S., M.S., and Ph.D. degrees from The Ohio State University, with his graduate training in ruminant nutrition. After earning his B.S. degree and before attending graduate school, Fox was a full-time crop and livestock farmer in Western Ohio. For the past 25 years, Fox's research has been focused on the nutrient requirements of cattle varying in biological type, and the development of computer programs to predict nutrient requirements and performance of cattle with wide variations in cattle type, feed composition, feeding system, environmental and management conditions. While at Cornell since 1977, he has conducted research in cattle nutrition, and he currently teaches a course on "Livestock and the Environment." Over the past 20 years, he and a team of scientists at Cornell have developed the Cornell Net Carbohydrate and Protein System for Evaluating Beef and Dairy Cattle Diets, which is widely distributed in the United States and internationally. Together with his colleagues, Fox has conducted pasture research for 15 years to evaluate pasture quality and matching cattle and forage management systems. In recent years, Fox has become involved in Cornell's Sustainable Agriculture program, and heads multidisciplinary projects on "Integrating Knowledge to Improve Dairy Farm Sustainability" and "Developing Software for Whole Dairy Farm Nutrient Management." His research and extension programs have resulted in over 150 invited presentations at conferences and symposia, over 500 research and extension publications, and 15 microcomputer programs. He has served on many national committees, including the National Integrated Resource Management Committee, and the National Research Council's Committees on Animal Nutrition and Feed Intake, and Subcommittee on Beef Cattle Nutrition.

JANE GOODALL is the world's foremost authority on chimpanzees, having closely observed their behavior for the past quarter century in the jungles of the Gombe National Park Game Reserve in Tanzania. Her observations and discoveries are internationally heralded. Her research and writing continue to make revolutionary inroads into scientific thinking regarding conservation and evolution. Goodall received her Ph.D. from Cambridge University in 1965. She has been the Scientific Director of the Gombe Stream Research Center since 1967. In 1984, she received the J. Paul Getty Wildlife Conservation Prize for "helping millions of people understand the importance of wildlife conservation to life on this planet." Her other awards and international recognitions fill pages. Goodall's scientific articles have appeared in many issues of National Geographic. She has written scores of papers for internationally known scientific journals. Goodall also has authored many books including *In The Shadow of Man* and *Through a Window*. Goodall attributes her dedication and insight to her work and her mission in life to her mother, internationally known author, Vanne Goodall. In 1985, Goodall's twenty-five years of anthropological and conservation research was published, helping us all to better understand the relationship between all creatures. She has now devoted over thirty years to her mission. Goodall expanded her global outreach with the founding of the Jane Goodall Institute in 1977, which is now based in Silver Spring, Maryland. She teaches and encourages young people to appreciate chimpanzees and all creatures great and small. Goodall lectures, writes, teaches and continues her mission in many inventive ways, including the Roots and Shoots environmental and humanitarian education program for young people.

DONALD B. JUMP is the Director of Research and Graduate Studies for the Department of Physiology at Michigan State University, East Lansing, Michigan. He is jointly appointed in the Department of Biochemistry and holds the rank of Professor in both the Physiology and Biochemistry Departments. He received his Ph.D. degree in biochemistry from Georgetown University in 1979. Afterward, he was a postdoctoral fellow with Jack Oppenheimer in the Endocrinology and Metabolism Section, Department of Medicine at the University of Minnesota in Minneapolis. He was appointed to assistant professor of medicine at the University of Minnesota in 1982. In 1985, Jump moved to the Physiology Department at Michigan State University. He has served on the editorial board for the Journal of Biological Chemistry and has served as an ad hoc reviewer for several NIH study sections and international granting agencies. Jump has co-chaired sessions at scientific meetings on nutrients and gene expression. He has authored more than 80 peer-reviewed journal articles, invited chapters, and reviews. His research has been funded by NIH, USDA, the American Diabetes Association, and the Michigan Agriculture Experiment Station. His research focuses on dietary fat regulation of gene transcription, with particular emphasis on fat effects on lipid metabolism in liver and white adipose tissue. His studies were the first to document fatty acid-regulated-*cis*-regulatory elements in genes encoding proteins involved in hepatic lipid synthesis.

KIRK C. KLASING is Professor of avian nutrition at the University of California, Davis. He received a B.S. degree at Purdue University and a Ph.D. at Cornell University in 1982. Since 1985, he has been in the Department of Avian Sciences at the University of California, Davis. His research interests include the interactions between nutrition and the immune system of animals, for which he has received the Poultry Science Research Award from the Poultry Science Association, the BioServ Award from the American Institute of Nutrition, and the Lilly Animal Scientist Award. Klasing serves on the editorial boards of Poultry Science, Animal Biotechnology, and Amino Acids. He is the author of a book on Comparative Avian Nutrition, as well as 75 refereed and 115 non-refereed articles and 6 book chapters.

QUINTON R. ROGERS serves as a Professor of physiological chemistry in the Department of Molecular Biosciences in the School of Veterinary Medicine at University of California, Davis. After receiving his B.S. degree in agriculture from University of Idaho in 1958, he completed an M.S. and a Ph.D. degree in biochemistry at University of Wisconsin, Madison by 1963. Following predoctoral and postdoctoral N.I.H. fellowships, Rogers progressed from the position of research associate to assistant professor of physiological chemistry in the Department of Nutrition and Food Science at Massachusetts Institute of Technology. In 1966, he was appointed assistant professor of physiological chemistry in the department where he currently serves, and by 1976 had achieved a full professorship. Rogers's research interests include biotechnology in nutrition and metabolism, experimental nutrition and metabolism of amino acids, control of food intake, feline and canine nutrition, and taurine nutrition. In 1986, Rogers received the Ralston Purina Small Animal Medicine Research Aware in Nutrition, and in 1992, the School of Veterinary Medicine at University of California, Davis conferred its Faculty Research Award on him. The American Society of Nutritional Sciences honored him with the Osborne Mendel Award. Rogers serves as an Honorary Diplomate of the American College of Veterinary Nutrition.

MICHELLE C. ROSS received the degree of Doctor of Veterinary Medicine and a Master of Science degree in physiology from Colorado State University in 1981. In 1995, she completed her Ph.D. in the physiology department of the John Burns School of Medicine at University of Hawaii. She practiced as a large animal veterinarian from 1981 to 1985. Since 1986, LTC

Ross has served in the United States Army Veterinary Corps, where she is chief of the Drug Assessment Division at U.S. Army Medical Research Institute of Chemical Defense in Aberdeen, Maryland. From 1995 to 1996, she acted as principal investigator, cardiac pathophysiology, for a project to assess clinical parameters of cardiac damage following nerve agent exposure. Previously, she served as director of preventive medicine and senior marine mammal veterinarian at the Naval Ocean Systems Center in Kailua, Hawaii from 1987 to 1992.

PHILIP A. THACKER was born in Vancouver and received his B.Sc. in 1974 and his M.Sc. in 1978, both from the University of British Columbia. He was awarded a Ph.D. in 1982 by the University of Alberta for his work on the effects of dietary propionate on lipid metabolism in growing swine. Thacker increased his awareness of the swine industry as a regional swine specialist with Alberta Agriculture after graduation. In 1984, he was appointed Assistant Professor in the Department of Animal Science at the University of Saskatchewan. He was promoted to Associate Professor in 1987 and Full Professor in 1991. Thacker is active in research, teaching, and extension. He is the author of over 110 refereed publications in scientific journals, including four scientific reviews. Other publications include conference proceedings, abstracts, and many technical reports. During his career, he has made 28 conference and 41 extension presentations. He co-authored a book on general swine nutrition for swine producers and edited a book on Non-Traditional Feed Sources for Use in Swine Production. Thacker's main areas of research center on evaluating new feed sources for use in swine production and in developing methods to increase the reproductive efficiency of the sow herd. He has evaluated the potential of alternative feeds such as buckwheat, hulless barley, rye, wild oat groats, and fish silage, as well as the potential to use enzymes to improve their value. In addition, he has conducted studies to determine the effectiveness of growth hormone, relaxin and gonadotropin-releasing hormone as a means of improving the reproductive performance of swine. Thacker is widely sought after as a speaker for swine extension meetings. He provides scientific information in an understandable format that is appreciated by swine producers across Canada. In addition, he served on the National Research Council's Committee on Animal Nutrition Subcommittee on Swine Nutrition, which recently published the tenth revised edition of *Nutrient Requirements of Swine*. He received the Young Scientist Award from the Canadian Society of Animal Science in 1989.

DUANE E. ULLREY is professor emeritus of animal science, fisheries, and wildlife at Michigan State University, and is Chair of the Committee on Animal Nutrition's Subcommittee on Nonhuman Primate Nutrition. He also serves as research associate for the Jennings Center for Zoological Medicine in San Diego, and the Smithsonian Institution's Department of Zoological Research at the National Zoological Park in Washington, D.C. Ullrey has devoted significant efforts throughout his career to improving dietary management for the betterment of animal health, welfare, and conservation of endangered species. His research interests include quantitative nutrient requirements of various domestic and wild species and analytical methods applicable to their study, mineral and vitamin metabolism, nutrition and immunologic response. As a professor and mentor of many students, Ullrey has contributed to the basis for education of many professionals throughout the world in nutrition. Ullrey has devoted almost two decades to National Research Council committee activities, as chair of the Committee on Animal Nutrition and as a member and chair of numerous subcommittees. He has also served on panels and committees for the National Science Foundation, the American Institute of Nutrition, the American Society of Animal Science, the National Institutes of Health, and the Smithsonian Institution.

DANIEL F. VILLAMAR holds a bachelor's degree in Agriculture (Food Science) from the University of Maryland, a master's degree in Biology from California State University, and a Ph.D. in Animal Science from Texas A&M University. His graduate research was focused on marine shrimp larval development and nutrition. Since joining Cargill, Villamar has received the Corporate Achiever's Circle Award, Cargill's highest honor for technical excellence, and has been granted a U.S. patent for developing LiquaLife®, the world's first liquid shrimp feed. Currently, Villamar leads the development of Cargill's aquaculture product line in the United States, Latin America, East Asia, and Eastern Europe with the deployment of Cargill AquaFeed™ products for finsfish and *Crustacea*. Before joining Cargill, Villamar worked in the U.S. feed industry as Research Manager, Research Scientist, and principal investigator on USDA, NSF, and privately-funded projects ranging from development of artificial kelp for abalone to more conventional pelleted, extruded and flake feeds.

BRUCE A. WATKINS is professor of food science and nutrition at Purdue University, and adjunct professor of anatomy in the Department of Anatomy, School of Medicine, Indiana University Purdue University Indianapolis. He received both his B.S. and M.S. degrees in nutrition from Colorado State University, and his Ph.D. degree in nutrition and physiological chemistry from the University of California, Davis, in 1985. He received the a National Research Award for his work on biotin metabolism in 1990, and in 1994 was presented the BioServ Award from the American Society of Nutritional Sciences (ASNS) for his research on the biochemistry of fatty acids in bone. His research interests include: food lipids, lipid biochemistry, eicosanoid and growth factor regulation of bone modeling, antioxidant nutrient interactions in chronic disease, plant phytochemicals, nutrient-gene regulation and molecular biology. Watkins is the author of more than 100 publications, which include refereed manuscripts, book chapters, and reviewed proceedings. He has given more than 30 national and international invited talks since 1991. Watkins serves on three editorial boards for nutrition and food science related journals. He is a Food Science Communicator for the Institute of Food Technologists and a member of the Guide of Experts in Lipid Metabolism for the ASNS and American Society for Clinical Nutrition. He teaches courses on lipid chemistry, nutritional sciences, and functional foods.

SUSAN YANOFF received her D.V.M. degree from Cornell University in 1980. After three years of private practice, she entered active duty in the Army Veterinary Corps. In 1991, she completed a residency in small animal surgery, as well as a master's degree, at Texas A&M University College of Veterinary Medicine. Yanoff is a Diplomate of the American College of Veterinary Surgeons and the American Board of Veterinary Practitioners. Her assignments include, Commander of the 51st Medical Detachment in Germany, Chief of Clinical Services at the Department of Defense Military Working Dog Veterinary Service at Lackland Air Force Base in San Antonio, and Commander of the National Capital District Veterinary Command at Ft. Belvoir, Virginia. She is currently stationed in Heidelberg, Germany as the Deputy Commander for Theater Support at the 100th Medical Detachment.

Appendix

SYMPOSIUM PROGRAM

Scientific Advances in Animal Nutrition: Promise for the Next Century

70th Anniversary National Research Council's Committee on Animal Nutrition
December 9, 1998

National Academy of Sciences Auditorium 2101 Constitution Avenue, N.W. Washington, D.C.

8:30 a.m.	**Registration**
9:00	Welcome and Opening Remarks
Dale E. Bauman
Chair, Symposium

Gary L. Cromwell
Chair, Committee on Animal Nutrition

Michael J. Phillips
Director, Board on Agriculture |
| 9:30 | Landmarks and Historic Contributions of Animal Nutrition
Duane E. Ullrey
Michigan State University and formerChair, Committee on Animal Nutrition |

10:00	**Keynote Address: Inroads to Animal Conservation** Jane Goodall *The Jane Goodall Institute*
11:15	**Recent Developments in Animal Nutrition** *Session Moderator: Donald C. Beitz* *Iowa State University*
11:20	Protecting Animal Health: Nutrition and Animal Immune Function *Kirk Klasing, University of California, Davis*
11:40	Designing Foods: Feeding Animals to Reduce Human Health Risks *Bruce Watkins, Purdue University*
12:00 p.m.	Metabolic Modifiers: Advances in Economic Production of Safe Food *Bob Collier, Monsanto*
12:20	Nutrients as Regulators of Gene Expression *Donald B. Jump, Michigan State University*
12:40	Discussion Session *Rapporteur: Karin Wittenberg, University of Mannitoba*
12:55	LUNCH
1:55	**Animal Nutrition's Crucial Role in Worldwide Endeavors** *Session Moderator: Mary E. Allen, Smithsonian Institution National Zoological Park*
2:00	Animal Nutrition in Space: Experiences of a Veterinary Shuttle Astronaut *Rick Linnehan, NASA (public appearance only)*
2:20	Our Changing Environment: Developing Strategies for the Future *Danny Fox, Cornell University*
2:40	Readiness of Military Service Animals *LTCOL Michelle Ross, U.S. DoD* *LTCOL Susan Yanoff, U.S. DoD*
3:10	BREAK

3:25	Research and Education Needs for the Next Generation *Quinton Rogers, University of California, Davis*
3:45	Discussion Session *Rapporteur: Michael Galyean, Texas Tech University*
4:00	**International Aspects of Animal Feeds, Feeding, and Nutrition** *Session Moderator: John Halver, University of Washington*
4:05	International Relevance of Feed Information *Philip Thacker* *University of Saskatchewan, Canada*
4:25	The International Market and Global Needs *Dan Villamar, Cargill*
4:45	Discussion Session *Rapporteur: Joe Fontenot, Virginia Polytechnic and State University*
5:00	**Finale: Implications of Committee on Animal Nutrition Studies for Meeting Challenges of the Next Century** *Dale E. Bauman* *Cornell University* *(Broadcast from Brussels, Belgium)*
5:20	Announcements and Closing Remarks *Gary L. Cromwell*
5:30 p.m.	RECEPTION *Great Hall* *National Academy of Sciences*